NATIONAL CITY PUBLIC LIBRARY

D0851997

Melting the Earth

Melting the Earth

The History of Ideas on Volcanic Eruptions

Haraldur Sigurdsson

New York Oxford
Oxford University Press
1999

Oxford University Press

Oxford New York
Athens Auckland Bangkok Bogotá Buenos Aires Calcutta
Cape Town Chennai Dar es Salaam Delhi Florence Hong Kong
Istanbul Karachi Kuala Lumpur Madrid Melbourne Mexico City
Mumbai Nairobi Paris São Paulo Singapore Taipei Tokyo
Toronto Warsaw

and associated companies in
Berlin Ibadan

Published by Oxford University Press, Inc.
198 Madison Avenue, New York, New York 10016

Library of Congress Cataloging-in-Publication Data
Sigurdsson, Haraldur.
Melting the earth : the history of ideas on volcanic eruptions / by Haraldur Sigurdsson.
p. cm. Includes index.
ISBN 0-19-510665-2
1. Volcanism. I. Title.
QE522.S548 1999
551.21–dc21 98-20299

Frontispiece: woodcut depicting eruption of Mount Etna above the walled-in-town of Catania; from *Cosmographia Universa* of Sebastian Munster, 1541 (author's collection).

Woodcut on page x by Anthony Russo (author's collection).

9 8 7 6 5 4 3 2 1

Printed in the United States of America
on acid-free paper

Contents

Preface

One thing I have learned in a long life: that all our science, measured against reality, is primitive and childlike—and yet it is the most precious thing we have.

Albert Einstein

The seed for the idea of this book was planted in 1979 when I was carrying out field studies of volcanic deposits from the famous A.D. 79 eruption of Vesuvius in Italy. In my attempts to integrate the evidence from the excavations in Pompeii and Herculaneum with the eyewitness report of Pliny the Younger, I began to read the accounts by the ancient Greek and Roman scholars of other volcanic eruptions.[1] What began as casual reading has gradually developed over the years into a fascination with the evolution of ideas about the causes of volcanic eruptions. In a way, my journey through these old sources has also been a very personal search for my intellectual roots, as a scientist who has devoted a career to the study of volcanic activity. It has been a rewarding journey, but also a somewhat humbling experience. As I have mapped out the history of knowledge of volcanic phenomena and traced the geologic thoughts of our predecessors, I have been amazed at how much they knew—and how early they knew it. I have been even more amazed at how some fundamental theories have been put forward, but then forgotten or ignored, only to be reinvented. It seems that many scientists consider that their field of science has always existed and give little or no thought to the long process that has gone before, during the evolution of their discipline over several centuries. Thus, Earth science professors do not dwell as a rule on the history of their science, but instead lecture to their students in an expository manner. I have also learned from working on this book that most scientists are guilty of arrogance toward early research—an arrogance that is built on ignorance. The editor of a scientific journal or the reviewer of a research paper invariably checks that the author is up to date, (i.e., has included references to the most recent publications), but what of the *earliest* publications on this topic? The scientist or scholar is the keeper of the flame of knowledge and he or she also advances our knowledge in a chosen field of research, but should also be responsible for linking the present with the past and maintaining a record of the history of knowledge in that field.

My fascination with volcanic activity dates back to my teens, when I was growing up in the volcanic regions of western Iceland. During what seemed at the time to be endless treks behind my father through the mountains to reach distant trout lakes and salmon rivers, I was continually exposed to barren, austere, but beautiful volcanic landscapes that early on aroused my curiosity about how the Earth works. Besides, at least two of my ancestors have also made a living from volcanoes or their products, although in a much more practical manner than I have in my academic career. My great-great-great-grandfather in a direct paternal lineage, Nikulás Buch (1754–1805), was in charge of the sulfur mine near Húsavík in north Iceland from 1786 to 1791, where the bright yellow sulfur encrustations were scraped off the surface of hot fumaroles. Nikulás was an immigrant to Iceland from Norway and is best known for introducing skiing to Iceland. My grandfather Steinthór Björnsson (1860–1926) made a living as a stone mason and split many a volcanic stone in Iceland, including the rocks that were used to build the House of Parliament in Reykjavík.

This is not a book written by a historian of science, but one by a practicing scientist who is deeply interested in the history of his science. It is about the manner in which magma or molten rock is generated at high temperatures within the Earth. It would seem logical that the history of ideas on melting in the Earth would be synonymous with the history of volcanology. This is not the case, however, because up to the mid-1970s volcanology was essentially a descriptive endeavor, primarily devoted to the geomorphology of volcanic landforms, the geography of volcanic regions, and the chronology of eruptions. Until recently, volcanologists have shown little insight into the physical processes of volcanic action and have often ignored the causes of the melting processes that lead to the formation of magma. Volcanology thus became a descriptive field, lacking rigor, and on the fringes of science, allied more with geography and geomorphology.

In the early part of the twentieth century a great effort began by petrologists and geochemists to apply chemistry and mineralogy to the study of the compositional variation of volcanic rocks and determine the source region of their origins. These studies were not, however, directed toward solving the riddle of how melting occurs in the Earth. As this book demonstrates, the processes that bring about melting in the Earth had been essentially discovered by physicists in the mid-nineteenth century, but due to the bifurcation of scientific fields, volcanologists were on the whole ignorant of these findings of the early natural philosophers or geophysicists and continued to pursue a descriptive approach of surface features without acquiring an understanding of (or seeming to be particularly curious about) the fundamental processes at work in the deep Earth. This bifurcation between volcanologists and geophysicists is one example of "the great barrier" that the Canadian geophysicist J. Tuzo Wilson recognized in 1954 as separating two great groups of Earth scientists.[2] Wilson pointed out that one group (the field geologists) deserves praise "for their assiduity, care, and often courage in collecting a monumental volume of data about those detailed surface features which are exposed for them to see." In contrast, another group (geophysicists and geochemists) was devoted to the study of the deeper processes, but had failed to understand them because they were "unable to analyze or make use of the field data." He considered it ironic that this group had a better understanding about "the nature of the nuclear processes which heat the sun and characterize the atoms of different elements" than about "those slow movements within the earth which have generated its powerful magnetic field, developed its mountains, and given structure to its crust."

I owe personal debts of gratitude to two Icelandic men who put me on a track to seeking knowledge. One is my late father, who by setting a fine example taught me to love nature, cherish books, and treat them with reverence. The other is the late geologist Sigurdur Thórarinsson, a true Renaissance man who blazed the way in exploring the impact of geologic processes on history and human economy. For inspiration on the history of volcanology, I am indebted to the book *Les feux de la terre* by my friend the late Maurice Krafft, who was killed during the 1991 Unzen eruption in Japan.[3] For the history of Earth science, I am especially indebted to the works of Stephen G. Brush,[4] the book *From Mineralogy to Geology* by Rachel Laudan,[5] and the groundbreaking Ph.D. thesis of David S. Kushner.[6] I must thank Roberta A. Doran of the Pell Library for locating rare and obscure texts worldwide, and Joyce Berry (my editor at Oxford University Press) and Rosaly Lopes-Gautier for valuable suggestions on editing the manuscript.

Jamestown, R.I.
July 1998

Haraldur Sigurdsson

Melting the Earth

1

The Heat Below

Just as the dairy-maid believes the moon to be a great cheese, so the astronomer fancies our globe a condensed nebula; the chemist, an oxydized ball of aluminium and potassium; the mineralogist, a prodigious crystal—"one entire chrysolite"; and the zoologist, an enormous animal—a thing of life and heat, with volcanoes for nostrils, lava for blood, and earthquakes for pulsations.

George Poulett Scrope, 1835

Heat and fire, bringing warmth and light to our planet, have been of crucial importance to humanity's progress. In prehistoric times, opportunities to "discover" fire and observe heat abounded. Lightning strikes ignited fires in the primeval forest or on arid grassy plains. Volcanic eruptions and lava flows threw forth showers of glowing sparks and incandescent heat. Most likely, these were the sources our ancestors borrowed from, carefully nurturing the fires they kindled, for these sources could be fickle and sporadic. With fire at their command, humans could venture into new realms of the cold and the dark—into polar terrain and mountain caves. Fire may also have contributed to a sense of community and language as our ancestors gathered around their hearths at night.

For *Homo erectus*, managing fire represented a great technological and cultural advance.[1] The earliest confirmed evidence of this comes from the Zhoukoudian cave in China, where 600,000-year-old hearths and several layers of charcoal testify to the use of fire by the hominid commonly called Peking Man. In 1993 burnt clay was found at an African site—Chesowanja in Kenya—a discovery that may push the date as far back as a million and a half years.[2]

Evidence of the early use of fire also comes from other

archaeological finds, such as stones that were fired before they were worked into tools. Several hundred thousand years ago people discovered that heating made stones more brittle and easier to fashion into tools, producing cleaner fracture surfaces, which in turn gave the tools a sharper edge. Similarly, the tips of wooden spears were frequently hardened by inserting them into fire.

Several legends about the origin of fire attribute it to a volcanic source or a place deep in the Earth. In the fire myth of the Polynesian people of the Tonga Islands, the great primeval hero Maui resides deep in the netherworld and is the keeper of fire. Maui, the legend goes, sleeps in a deep cave and, when he turns while dreaming, an earthquake occurs on the land above. In the Samoan Islands dwelt the earthquake god Mafuie, who tended eternally burning fires in fiery subterranean ovens. Mafuie was said to blow into the oven, scattering hot stones and cinders about—a vivid evocation of a volcanic eruption. In the Marquesas Islands, the goddess of fire and earthquakes was Mauike, who lived in the netherworld Havaiki and gave fire to the human race. The Athapascan Indians of British Columbia tell a story of the origin of fire, in which a column of smoke and flames, then tongues of fire, issued from a great mountain on the horizon, a myth suggesting one of the active volcanoes of the Cascades as the source.[3]

Our knowledge of human precursors and early humans is also, in a curious way, intimately liked with volcanic eruptions. Because of the excellent preservation of fossils in volcanic deposits, paleoanthropologists have begun to unravel the mystery of human origins. Most of the oldest remains of early humans come from volcanic regions in Africa and Indonesia and the association of volcanic activity with these fossils is no mere coincidence. The most logical explanation for their relative abundance and good preservation in such environments is that the bones were rapidly covered by volcanic deposits.

The degree of preservation is almost entirely dependent on the sedimentation rate —the rate at which deposits accumulate on top of bones. This process is relatively rapid on the ocean floor, but of course this is not the realm for our fossil search. On dry land, however, the deposition of sediment is generally no faster than the rate at which soil forms and so human remains are normally exposed to scavengers, weathering, and erosion for decades before they are buried and sealed into the soil layer. Thus, bones may simply vanish during their long exposure. In volcanic regions, however, deposits of pumice, ash, lava, mudflows, and other sedimentary products resulting directly or indirectly from eruptions accumulate very rapidly. A sudden fall of ash and pumice can deposit a blanket ranging from a few inches to several feet in thickness in a single day, covering thousands of square miles. Just as the Roman cities of Pompeii and Herculaneum were instantly buried by the eruption of Vesuvius in A.D. 79 and preserved virtually intact, so have earlier volcanic deposits sealed in the remains and fragile artifacts of our more distant ancestors. As Mary Leakey has shown, the stunning discovery of 3.7-million-year-old *Australopithecus* hominid footprints crisscrossing a volcanic ash deposit at Laetoli is a graphic testament to the power of preservation of volcanic ashfall.[4]

Tectonic movements may also accompany volcanic activity, often with rifting of the

Earth's crust, leading to rapid uplift of the adjacent land and sinking of the rift valley. These uplifted regions are then subjected to intense erosion, with sediments deposited in lakes that form within the valley. One region where rifting and volcanic activity has beautifully preserved hominid remains is the Olduvai Gorge, within reach of the great Ngorongoro crater in a valley known as the East African Rift. Here the fossil-bearing sediments are shallow-water lake deposits of volcanic material, interbedded with volcanic ash layers. Volcanic deposits also contain minerals that can be conveniently dated by measuring the decay of the radioactive isotopes potassium and argon, a method that has enabled the dating of fossil remains of the oldest human, *Homo habilis*, at Olduvai as 1.75 million years old. A similar setting in the African rift at Hadar in Ethiopia has yielded fossils of a possible human precursor, *Australopithecus afarensis*, permitting its dating at 3.5 million years.[5] Is the relative abundance of hominid and early human remains in volcanic regions simply a preservation factor or is it perhaps also related to the preference of our ancestors for areas of active volcanism? A case could be made for such attraction—humans in search of fire or drawn to the highly fertile soils and abundant game of volcanic areas.

Volcanic stone is the earliest known material used in the creation of tools by the genus *Homo*. The most primitive stone artifacts, implements made from lava, are about 2.5 million years old, and were found at Lokalelei, on the shore of Lake Turkana in East Africa.[6] Following their humble beginnings as toolmakers with the fashioning of these crude "cores," early humans gradually improved the techniques of working stone into flakes, choppers, hand axes, cleavers, and, finally, delicate obsidian stone blades in the upper Paleolithic (Stone Age). Obsidian, which became the raw material of choice for early toolmakers, is produced when magma, or molten rock, flows out of a crater and quickly solidifies to form a black natural glass of great hardness and beauty. It has the advantage of being extremely hard and durable, but also massive. Free of joints and cracks, it splits with a clean fracture when struck in an appropriate manner, and could be fashioned into needle-sharp points and blades as keen as a surgeon's knife. For millennia, volcanic regions in Africa, the Mediterranean, and Central America have served as the world's principal source of obsidian.

More mundane materials that are the characteristic products of explosive volcanic eruptions are pumice and volcanic ash—the latter has been used for nearly 3000 years in making cement. A mixture of volcanic ash and lime produces a very durable and water-resistant (hydraulic) cement.

Probably the oldest example of its use, dating from the seventh century B.C., comes from the lining of a cistern on the Aegean island of Rhodes.[7] In Roman times, ash was taken from the volcanic deposits near the port of Pozzuoli in the Bay of Naples and the cement became known as *pozzolana*. The Romans put pozzolana to good use in the construction of the Pantheon (ca. A.D. 10) and the Colosseum (ca. A.D. 70). They even employed the cement in the underwater building of Roman harbors, including their major port at Ostia. Use of this valuable raw material continued even after the invention of Portland cement in the nineteenth century, and as late as the 1860s pozzolana from volcanic ash deposits on the Mediterranean island of Thera was used in constructing the Suez Canal.

Volcanoes are also a principal source of sulfur, once known as brimstone. Early on, people realized that sulfur burns with a strange and sputtering flame, giving off an evil-smelling and choking vapor. Probably the first uses for sulfur were to kill insects and bleach wool, feathers, and fur. Homer knew of its fumigation properties and praised the "pest-averting sulfur." Ulysses calls out: "Quickly, bring fire that I may burn sulfur, the cure of all ills." Over the centuries the burning of a sulfur candle was a housecleaning ritual, following a case of contagious disease in the home. Sulfur features in Egyptian prescriptions of the sixteenth century B.C., and Pliny the Elder describes its medicinal, industrial, and artisan uses in his *Historia Naturalis*. (A.D. 77). The Romans also found a new application in warfare for the sulfur they mined in Sicily. By mixing it with tar, rosin, and bitumen, they produced the first incendiary weapon. Yet it was not until the invention of gunpowder in the thirteenth century that the demand for sulfur became widespread. The explosive formula of 70% saltpeter (sodium nitrate), 15% charcoal, and 15% sulfur was discovered by the Chinese and introduced to Europe by Roger Bacon in the early fourteenth century with revolutionary impact — helping to blast apart the feudal system and making armored knights history.

At times, the pressing need for sulfur has led individuals to take great risks to obtain it. During the Spanish conquest of the Aztecs in Mexico, Hernando Cortes renewed his gunpowder stock by ordering his soldiers lowered into the active Popocateptl volcano so they could scrape sulfur encrustations from its walls. Just climbing this 5452-m (17,883 ft) volcano is no small feat, let alone extracting 140 kg (300 lbs) of sulfur from its active crater. Partway up the volcano, after an arduous climb, the soldiers dug a cave in the snow and bivouacked there, but were driven from their shelter by sulfur fumes and the cold. When they resumed their ascent, an eruption shook the mountain and as they approached the rim of the crater, there was another explosion, but they remained unscathed. When the fumes and smoke cleared, the view into the crater revealed seething masses of lava. The men drew lots to determine who should make the first descent, and the lot fell to their leader Montano. His companions lowered him into the abyss on a rope tied around his waist, a distance of about 200 meters (650 ft). Fortunately the rope held, and he was raised and lowered seven times, each time delivering a bag of sulfur. The men then started their descent down the ice-covered flank of the volcano to be greeted as heroes when they reached the foot. Cortes welcomed them with an embrace, and, in a letter to the King of Spain (1524), described the sulfur mining venture, adding, "In the future this method of procuring it will be unnecessary; it is certainly dangerous and I am continually writing to Spain to provide us."[8]

Volcanoes are also the source of gold, silver, and diamonds. Such treasures are not found scattered about on the surface of active volcanoes, but in the exposed roots of ancient volcanoes. Yet it is not the practical aspects of volcanic eruptions that has led to our fascination with them, but rather that they represent the most awesome and powerful display of nature's force. The idea that *terra firma* may explode under our feet and bombard us with glowing hot ejecta seems almost incomprehensible. Every year about fifty volcanoes throughout the world are active above sea level, threatening the lives and property of millions. A single eruption can claim thousands of lives in an instant, as during the 1902 eruption of Mount Peleé on the Caribbean island of Martinique, when a flow of hot ash and gases overwhelmed the city of Saint Pierre, killing all but one of its 28,000 inhabitants. More recently, a mudflow triggered by the 1985 eruption of the volcano Nevado del Ruiz in Colombia killed nearly all of the 25,000 inhabitants of the town of Armero. Despite their awe-inspiring, spectacular, and even deadly fireworks, volcanic eruptions are not nature's most deadly hazard. The principal volcanic disasters since 1700 have taken 260,000 lives, which pales in comparison with the death toll from earthquakes and tropical storms. The deadliest disaster known was the Huahsien earthquake in China (1556), which killed more than 820,000 people. In recent times, the magnitude 7.8 Tangshan earthquake, also in China (1976), killed more than 240,000 people. The worst hurricane in history occurred in 1970 in the Ganges Delta in Bangladesh, with a loss of 500,000 lives. Yet living with a dormant or active volcano, looming large above a village or town, is a visible and permanent threat and qualitatively different from the transient threat of a hurricane or a sudden earthquake. As a source of limitless power, volcanoes have become the subject of superstition and generated legends, inspired religious worship, and even provided a logical geographic locus for Hell. Unwittingly, however, humans have been drawn to volcanic regions throughout the ages because of the extremely fertile soils produced by the disintegration of volcanic ash. In modern times, abundant geothermal energy from the roots of volcanoes has also set people on a new course of wrestling wealth from this capricious giant.

Throughout the ages people have struggled with the puzzle: *why* do volcanoes erupt? Underlying this is an even more fundamental question and a central theme of this book: what causes the Earth's rocks to melt?

The earliest known ideas about the cause of volcanic eruptions date to the Greek natural philosophers of the fifth century B.C. Anaxagoras proposed that eruptions were caused by great winds stored inside the Earth. When these winds were forced through narrow passages or emerged from openings in the Earth's crust, the friction between the compressed air and the surrounding rocks generated great heat, leading to melting of the rocks and the formation of magma. To anyone who has observed an explosive volcanic eruption, this is a perfectly logical idea, one that in fact was taken up by Aristotle and passed on by scholars until the Middle Ages. Initially the Roman philosophers adopted this Greek view of the causes of volcanic eruptions, but eventually they proposed another explanation, one more in keeping with the practical Roman mind. Thus, by the first century A.D., Seneca saw volcanoes as giant furnaces where the combustion of fossil fuels, such as coal, bitumen, or sulfur, was the driving force of eruptions. Seneca's ideas were remarkably long-lived and formed the basis of the interpretation of the causes of volcanic eruptions throughout the Middle Ages and even well into the eighteenth century. Study of the Earth, like many scholarly activities, suffered a setback with the growth of the new Christian religion, and the only role of volcanoes in this new world order was to serve as a reminder of the hellfires burning below. A new view finally emerged in the early seventeenth century, when the French philosopher René Descartes advanced the idea that the Earth was once a hot star like the Sun, only smaller and therefore much cooler. Volcanic eruptions he regarded as a residue of the vast internal heat of the cooling star. This concept of a vestigial heat in the deep Earth prevailed with many scholars until late in the nineteenth century and formed the basis of the influential view of nineteenth-century British physicist Lord Kelvin concerning the thermal history and age of our planet.

The experiments of medieval alchemists and early chemists led to another highly original idea about the fundamental causes of volcanism and the origin of heat in the Earth. It was probably Robert Hooke who first suggested in the seventeenth century that the heat given off during chemical reactions, such as the combining of iron with sulfur compounds, could account for heat in the Earth and volcanic eruptions. The chemical theory of terrestrial heat and volcanism reached its peak in the first years of the nineteenth century in the works of the English chemist Humphry Davy, following his discovery of the alkali earth elements and their exothermic (heat-giving) chemical reactions. Soon, however, it proved a dead-end hypothesis.

Although no progress was made during the eighteenth century in understanding the underlying causes of volcanism, a revolutionary stride was made in recognizing ancient basaltic rocks as being of volcanic origin, even though they might be far removed from active volcanoes. A heated debate took place between scholars who believed that these rocks were deposited as sediments in the ocean (the Neptunists) and others who claimed that they were of volcanic origin (the Plutonists). The victory of the Plutonists settled the volcanic origin of basalt, led to recognition of ancient lava flows, and fostered the beginnings of experimental petrology—the study of the melting and crystallization behavior of rocks at high temperatures and pressures. It was the pioneering work of the Frenchmen Guettard and Desmarest in the mid-eighteenth century that resulted in the recognition of ancient volcanoes and the volcanic origin of basalt and thus set the foundations for volcanology. This debate is the favorite theme of most historians of the science of volcanology, but, in my view, its importance for the evolution of ideas about volcanism and melting in the Earth has been exaggerated. Because this field of research developed as a branch of geology, its workers failed to address the fundamental geophysical questions regarding the origin of heat in the Earth and how rocks melt.

The major revolution in understanding melting in the Earth and the origin of volcanic rocks did not emerge from geology, but can be largely traced to the work of pioneer geophysicists in the mid-nineteenth century, such as the Englishman William Hopkins, who were well versed in the emerging science of thermodynamics—the study of the fundamental laws of energy and heat. The discovery of how the Earth melts to produce magma was made in 1837, when it was realized that rocks melt simply due to the decrease in pressure within the Earth. The first to propose that rocks melted at lower temperature as the depth or pressure decreased was the Scottish professor of mathematics John Playfair in 1802. The experimental verification of this relationship was carried out by William Hopkins, Gustav Bischof, and Robert Bunsen from 1837 to 1850. The true heroes of this story are two scholars who have generally not been regarded as pioneers in Earth science: George Poulett Scrope and William

Hopkins. They were the first to recognize the basic process of melting in response to decrease of pressure. Rocks will melt spontaneously, without addition of heat, if they move only slightly upward, to a region of lower pressure or closer to the Earth's surface. But how could rocks move upward in the Earth? Over a period of almost a century, scientists proposed many imaginative schemes. The solution turned out to be the generation of large-scale and very slow-moving currents within the deep Earth that transported rocks upward and brought about melting due to the decreasing pressure. Convective flow in the Earth had been first proposed in 1797 by Count Rumford, and the idea was later developed by the British geophysicists William Hopkins and Osmond Fisher.[9] It was Arthur Holmes who in the early years of the twentieth century finally brought together both the elements of melting and convection to account for the generation of magmas and origins of volcanism in the Earth. His ideas were the true foundation of plate tectonics. In 1904 the monumental discovery was made that the radioactive decay of certain chemical elements within the Earth provides the heat source to drive the convection currents. Since that time, we have merely been fine-tuning the theories that explain melting in the Earth.

Why Do We Study Volcanoes?

In the spring of 1816 the English poet Percy Bysshe Shelley and his eighteen-year-old bride, Mary Shelley, visited Switzerland, intending to vacation on the shores of Lake Geneva with the poet Lord Byron. Yet their summer plans were upset by a sharp climate change brought about by a volcanic eruption on the other side of the world. "At first we spent our pleasant hours on the lake, and wandering on its shores" wrote Mary Shelley.[10] "But it proved a wet, uncongenial summer, and incessant rain often confined us for days." Housebound because of the miserable weather, they sought other forms of entertainment. "We will each write a ghost story" said Lord Byron, and in this literary contest between Byron, the Shelleys, and Byron's personal doctor, John Polidori, Mary Shelley's monstrous masterpiece *Frankenstein* was born.

Strange atmospheric phenomena had been observed in Europe before the summer of 1816. In May 1815 the sunsets in England were exceptionally colorful, and by September the sky seemed to be on fire every evening. But the spectacular skies brought bad weather the following year, and the year 1816 is the worst on record in Europe, with the mean temperature about 3°C (37.4°F) below average across the entire continent from Britain in the north to Tunisia in the south.[11] In England the summer was miserable, with the mean temperature for June the lowest on record, or 12.9°C (55.2°F).[12] In France it was so cold that the wine harvest was later than at any time since record keeping began—delayed until November in those districts where the grapes were not already frozen on the vine.[13] In Germany crops failed, exacerbated by the war with Napoleon, and famine and price inflation were widespread.[14] The food shortage was so severe that Johann H. F. Autenrieth, professor of medicine in Tübingen, published a book titled *Introduction to the Baking of Bread from Sawdust*. In it he documents how his unfortunate dog increased steadily in weight from eating bread baked from the birch sawdust. Conditions in Germany became so intolerable that the people rioted in many districts and set off in search of better living conditions, starting the great migration from northern Europe east to Russia and west to America. The severe weather led to major crop failure in Europe and a doubling of the price of wheat from 1815 to 1817. When we consider that in the early part of the nineteenth century, before the Industrial Revolution, agriculture was the mainstay of society, it is clear that the climate change had a strong impact on the economy of the Western world.

In North America the year 1816 was also unusually cold and dry, especially in New England, and was known as "the year without summer." In early May most farmers had planted their corn, the main crop at that time. But by mid-May the weather deteriorated, with severe frost that froze the fields as far south as New Jersey.[15] As late as mid-June, frosts also plagued New England, New Jersey, and New York. Then a great drought began, stunting or destroying the few remaining crops. In early July frost again swept over New England, from Maine to Connecticut and again in late August in Maine, Vermont, and south to Massachussetts. By the end of September a final frost destroyed the few crops that had survived the frigid summer. These crop failures, both in Europe and North America, have rightfully been called "the last great subsistence crisis of the western world."[16] The growing season, of course, was much shorter than normal. In New England the average growing season is about 120 to 160 days, but in 1816 it was only 70 days in Maine, 75 in New Hampshire, and 80 in Massachussetts—a reduction of 40 to 50% (Figure 1.1). Many farmers suffered such heavy losses that they abandoned their farms and migrated to the western territories.

This severe but short-lived climate change was caused by the great 1815 eruption of the volcano Tambora on the Indonesian island of Sumbawa, which emitted several million tons of sulfur dioxide gas into the Earth's stratosphere. The gas quickly condensed to form an aerosol, a cloud of sulfuric acid droplets that enveloped the Earth and scattered solar radiation back to space, leading to a sudden cooling of the Earth's entire surface. Yet the connection between these climate phenomena in the Western world and the eruption of Tambora was not realized until almost eighty years later.

The idea that volcanic eruptions may influence climate can be traced back at least two millennia to the days of the Roman Empire. In 44 B.C. Mount Etna on Sicily erupted, sending up a haze that markedly diminished sunlight in Rome. Curiously, the event coincided with the assassination of Julius Caesar and, in the minds of many Romans, the death of their emperor was directly related to the eruption.[17] Plutarch writes that the haze rendered the rays of the sun so faint and it became so cold that fruit could not ripen and withered on the trees. Several other sources indicate that the climate was unusually severe around the Mediterranean in 44 and 43 B.C. in the wake of the eruption. In Egypt crops failed, resulting in famine, and wheat shortages in Rome. The crop failure in Italy was so severe that the Senate appointed prominent leaders such as Cassius and Brutus to direct the harvesting and trade of wheat. We do not know the volume of the volcanic aerosol cloud that Etna produced, but we do know that this volcano is one of the largest emitters of sulfur dioxide to the atmosphere today.[18]

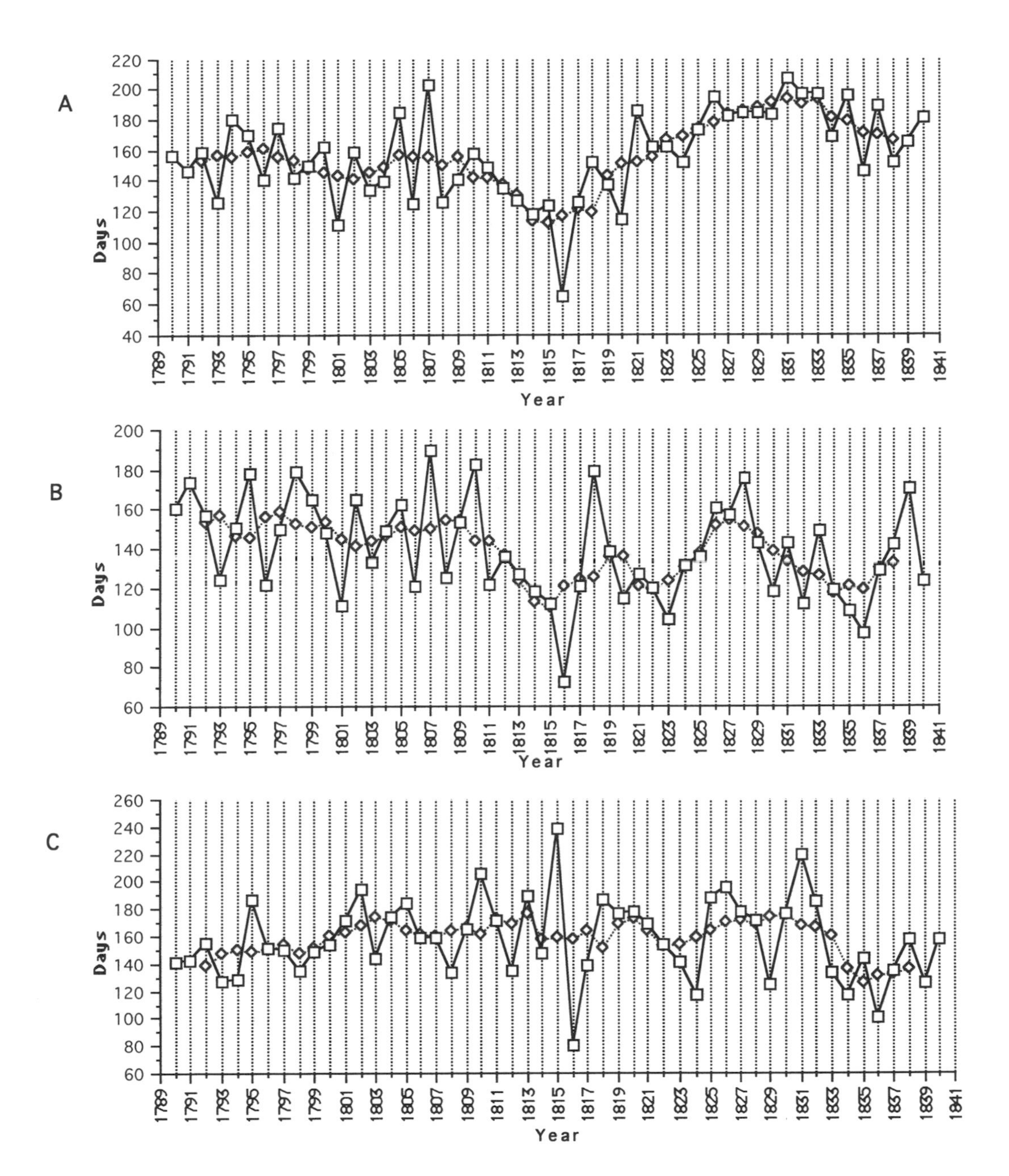

1.1. Volcanic eruptions can alter the climate. The great Tambora volcanic eruption in Indonesia in 1815 led to a short-term global climate change that dramatically shortened the growing season in New England the following year. The sulfur dioxide gas emitted by the volcano produced a worldwide veil of sulfuric acid aerosol in the stratosphere, decreasing the amount of solar radiation reaching the Earth's surface and causing global cooling. This figure shows the length of the growing season on an annual basis (dots) from 1789 to 1841, and the five-year running average (solid line). Panel A is for the state of Maine, B for New Hampshire, and C is the growing season in Massachussetts. Reproduced from C. R. Harington (ed.), *The Year Without Summer? World Climate in 1816* (Ottawa: Canada Museum of Nature, 1992).

The first scientific observations on the effects of volcanism were made in the wake of the Laki eruption in Iceland in 1783, when the volcanic haze, carried to the southeast, spread over Europe and across the Atlantic Ocean.[19] That summer Benjamin Franklin, residing in France as the American ambassador to the court of King Louis XIV, noticed that a persistent, dense haze lay over the land, dry and quite unlike ordinary fog. It so attenuated the strength of the Sun's rays that he could barely kindle brown paper with solar rays using a magnifying glass. Concluding that the ability of the Sun to heat the Earth that summer had been greatly diminished, Franklin proposed that the source of the haze could be traced to the volcanic eruption in Iceland.[20] Thus were laid the foundations for the study of the effects of volcanism on climate. Interest in how volcanism can affect the climate was rekindled by the 1815 eruption of Tambora. The American medical doctor Thomas D. Mitchell described the haze over New York in the spring and summer of 1816: "It had nothing of the nature of a humid fog. It was like that smoking vapour which overspread Europe about thirty years ago. The learned, who made experiments to ascertain its nature, could only state its remarkable dryness, by which no polished surface or mirror could be obscured." Here he is referring to Franklin's earlier observations. Although the haze and other climate phenomena were linked to volcanic activity, they were not attributed to Tambora until volcanological research was stimulated again after another great eruption, that of the Indonesian island of Krakatau in 1883.

The global climate impact of large eruptions is a major reason why an understanding of volcanic phenomena is important for society. A Tambora-size eruption in the twenty-first century will have much more profound effects on humans on an overcrowded Earth, where natural resources are strained to the limit. But climate change is but one of the many effects of volcanic eruptions. We are much more familiar with the local results of explosive eruptions, which devastate towns and villages by ash fallout and glowing hot ash flows. Such hazards have in recent years led to the development of volcano monitoring and prediction of eruptions.

2

From the Stone Age to Volcano Myths

The nether regions of the Earth are inaccessible in the ordinary sense. Before the time of Newton, when evidence about them was nearly totally lacking, it was not necessarily unreasonable to describe the Earth in terms of models involving say a Hell, or a subterranean monster shaking itself to cause earthquakes. The subsequent growth of evidence has lowered the plausibility of such models.

K. E. Bullen, 1975

Written records of ideas regarding volcanoes do not reach beyond the fifth century B.C., but legends and art come down to us from ancient times that give insight into earlier thought about volcanic activity. The earliest visual representation of volcanic eruption dates to one of the oldest village cultures on Earth, one that thrived on the Anatolian Plateau in southeastern Turkey more than 8,000 years ago. The site, known as Çatal Hüyük, was painstakingly excavated by the British archaeologist James Mellaart and his collaborators from 1961 to 1963.[1] Their work revealed a Stone Age town, with a long and rich cultural and artistic history, that flourished between 6500 and 5700 B.C. Numerous shrines found at the site attest to the development of religion and display a wealth of decoration, reliefs, and wall paintings. In their work the early artisans of Çatal Hüyük used rich colors and had a full range of pigments at their disposal, derived from such minerals as iron oxides, copper ores, mercury oxides, manganese, and lead. They also manufactured such artifacts as highly polished volcanic obsidian mirrors, ceremonial daggers, and metal trinkets—the latter evidence of their ability to smelt copper and lead.

Of particular interest is a wall painting that Mellaart found in one of the older shrines from a level that has been carbon-14 dated to soon after 6200 B.C. (Figure 2.1). In the foreground is a group of domino-like rectangular houses, which probably represents the town iteslf, and in the background rises a twin-peaked volcano, ochre in color and spotted with black dots, with an eruption spewing from the higher cone. This eruption is portrayed as curved lines or spray above the crater, and the dots on the cone and outside its outline may represent falling volcanic "bombs" or large pieces of semiliquid lava. The lines emerging from the base of the volcano may signify lava flows. Mellaart describes the painting thus: "The landscape painting from the north and east walls of shrine VII.14 shows a town in the foreground, rising in graded terraces closely packed with rectangular houses. An erupting volcano is shown behind the town, its sides covered with volcanic bombs rolling down the mountain slope. Others are thrown up from the erupting cone above which hovers a cloud of smoke and ashes. The twin cones suggest that it represents an eruption of Hasan Dag."[2]

Today the plain surrounding Çatal Hüyük is indeed dominated by the twin volcanic peaks of Hasan Dag and Karaca Dag and it is very likely that this ancient painting represents an eruption of Hasan Dag. It has been proposed that the painting had religious meaning, but several factors may have motivated the artist; perhaps it was a response to the threat from the erupting volcano, a work to appease the powers of the underworld. Was it inspired by gratitude to the volcano, the source of valued obsidian? A large number of obsidian spearheads and other implements have been found in the excavations, and it has been suggested that the town derived much of its wealth from the obsidian trade. Or was the painting simply inspired by fear in witnessing Earth's greatest display of power?

Volcanic activity in Anatolia also occurred in Roman times. Roman coins from Cappadocia portray an eruption of the stratovolcano Érciyas Dagi, an eruption referred to by Strabo the Geographer (63 B.C.–A.D. 20). Volcanic eruptions in Anatolia and other parts of

2.1. The oldest known picture of an erupting volcano dated about 6200 B.C. This painting in a shrine in the Stone Age town of Çatal Hüyük in Anatolia (eastern Turkey) shows an erupting volcano above the town whose houses are depicted as domino-like structures. According to the James Mellaart, the pattern on the erupting volcano indicates incandescent bombs rolling down the slope of the mountain while a plume of ash and gases hovers above. Reproduced from J. Mellaart, *The Earliest Civilizations of the Near East*, (New York: McGraw-Hill, 1965).

Turkey are a result of the shifting and interaction of the great African and Eurasian plates of the Earth's crust, which also account for the many deadly earthquakes over the centuries in this region. During millions of years, the African plate has been moving slowly but inexorably northward, steadily shrinking the Mediterranean basin. Consequently, the floor of the Mediterranean Sea has been thrust at an oblique angle under the Eurasian plate to the north, which includes Turkey and Asia Minor. This process of convergence and underthrusting of plates, referred to as subduction by geologists, leads to the formation of magmas or molten rock in the region of the Earth's mantle sandwiched between the two plates. Being hot and less dense than the surrounding mantle, the magmas rise and erupt at the surface to form volcanoes. In this manner, the volcanic islands of the Aegean Sea were created, including Santorini or Thera, where a famous Bronze Age eruption (1650 B.C.) may have precipitated the decline of the Minoan civilization.

Volcanic eruptions, the most spectacular and awe-inspiring of all natural phenomena, have throughout history aroused fear or curiosity, inspired religious worship, led to the creation of myths—and ultimately inspired scientific research. This list is not necessarily in chronological order because even in the twentieth century many, even all, of these responses can be observed when a volcano erupts. Today, when Mount Etna in Sicily erupts, villagers appeal to the Deity and bring out the sacred veil of their patron Saint Agatha to arrest the progress of lava flows that may be threatening their towns and vineyards. The Scottish medical doctor James Hutton, regarded by many as the father of geology, was in 1788 the first to attempt to explode the myth associated with volcanic eruptions: "A volcano is not made on purpose to frighten superstitious people into fits of piety nor to overwhelm cities with destruction."

Primitive people the world over have long believed that volcanoes were inhabited by deities or demons who were highly temperamental, dangerous, and unpredictable. To appease these capricious gods, humans have for centuries made the ultimate sacrifice. Thus, the Mayas, Aztecs, and Incas offered humans to their volcanoes.[3] Nicaraguans long believed that their dangerous volcano Coseguina would stay quiet only if a child were thrown into the crater every twenty-five years.[4] Similarly, young women were thrown into the crater of Masaya volcano in Nicaragua to appease its fire.[5] Until recently, people on the Indonesian island of Java sacrificed humans on Bromo volcano; they still throw live chickens into the crater once a year. People living near the feared volcanoes of Nyamuragira and Niyragongo, north of Lake Kivu in central Africa, sacrificed annually ten of their finest warriors to the cruel volcano god Nyudadagora. To those who were skeptical of such rites and pointed out that earlier sacrifices had failed to prevent or stop an eruption, believers have countered with the argument that things would have been much worse without the sacrifice.

The Aztec people named the volcanoes surrounding the Valley of Mexico after their gods. Popocatepetl and Iztaccihuatl, lying to the east of the valley, were worshiped as deities, linked to a beautiful love story. The enormous (5452 m; 17,882 ft) and frequently active volcano Popocatepetl (smoking mountain) was, according to Aztec legend, a warrior who fell in

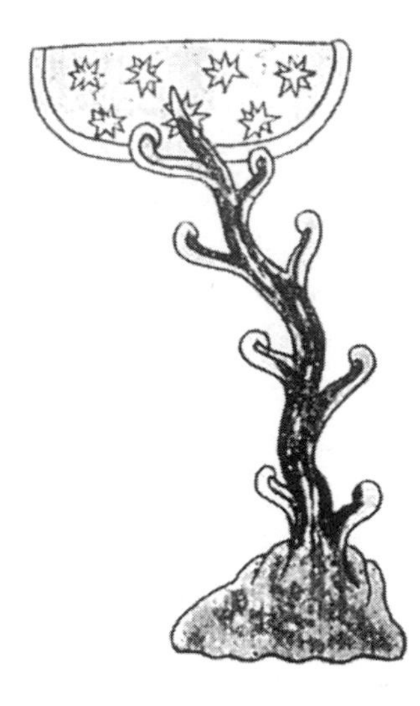

2.2. A drawing of a volcanic eruption, with the smoke and flames reaching the stars. From the *Codex Telleriano Remensis,* a Mexican manuscript dating from 1509. Reproduced from S. N. Coleman, *Volcanoes New and Old,* (New York: The John Day Company, 1946).

love with nearby Iztaccihuatl (sleeping woman), the daughter of the emperor.[6] When Popocatepetl returned victorious from war to claim his beloved, his enemies sent word ahead that he had been killed and Princess Iztaccihuatl died of grief. Popocatepetl then built two great mountains. On one he placed the body of Iztaccihuatl; on the other he stands eternally, holding her funeral torch. In an old Mexican manuscript, a volcanic eruption is documented and illustrated: "In the year 1500 on a very clear night which lasted forty days, there were those who thought that they saw a New Spain which was very large and resplendent, which was in the direction of the East and rose from the earth and reached the sky." In the text is a drawing of a volcanic cone, with a red and yellow plume reaching the starry sky (Figure 2.2). "This was one of the marvels which they saw. They thought it was Quetzacoatl [the Mexican god] for whom they were waiting."[7]

Volcanic eruptions featured prominently in Greek mythology, which abounds with allusions to volcanoes, associating them with such gods as Pluto, Persephone, Vulcan, and the fearsome Typhon. The idea that volcanic activity represents the stirrings of the Titans, giants imprisoned in the Earth, goes back to the classical time of the Greeks and perhaps further. Mott Greene has proposed that the mythical association of volcanic eruptions with battles between the Olympian gods and the Titans very likely dates back to preclassical antiquity. The Greeks saw the Titans as huge man-creatures, born of the Earth to attack the gods. Confined under various volcanic regions, their appearance on the land or in the air seemed the logical precursor to a volcanic eruption. The most awe-inspiring of all the Greek monsters was Typhon. The firstborn of Gaia (Mother Earth) and Zeus, he was the largest monster that ever lived. His arms, when spread out, reached a hundred leagues, fire flashed from his eyes, and flaming rocks hurtled from his mouth.[8] Even the gods of Olympus fled in terror at his sight. Typhon rebelled against the gods and even opposed Zeus, who then hurled Mount Etna at him, trapping the frightful creature under the mountain. When Typhon was thus imprisoned, a hundred dragon heads sprang from his shoulders, with eyes that erupted flames, a black

2.3. According to Greek mythology, the eruptions of Mount Etna in Sicily are caused by the stirring of the giant Typhon (known as Enceladus to the Romans), trapped underneath the volcano. When Typhon becomes restless, Zeus launches thunderbolts down to Earth to keep the monster under control. This engraving showing Typhon under Etna is from *The Temple of the Muses: Or the Principal Histories of Fabulous Antiquity* (Amsterdam: Zachariah Chatelain, 1733) in author's collection.

tongue, and a terrible voice. Each time Typhon stirs or rolls over in his prison, Etna growls and the Earth quakes, with eruptions and veils of smoke covering the sky (Figure 2.3). Typhon was to become the personification of evil, corresponding to the Egyptian god Set and the Romans Enceladus.

The jailkeeper of Typhon under Mount Etna was Hephaestus, the Greek volcano and fire god, who was later equated with the Roman god Vulcan.[9] Hephaestus was born weak and ugly to Hera (wife of Zeus), who, disappointed, threw him down to Earth from Olympus. On Earth the cripple Hephaestus learned the trade of metalworking and became a great artist; he was given the goddess of beauty and love Aphrodite as his wife, much to her disgust. His fabled forges were to be found under several Mediterranean volcanoes, including Vulcano, Etna, and Thera (Santorini), where he was assisted in his metalworking by the Cyclopes, one-eyed creatures whose hammer blows could be heard in the rumblings of the volcanoes. The legendary association of the Cyclopes with the Etna volcano probably stems from the glow in the night sky over the summit crater, visible far away as a red "eye" reflected from the clouds back to Earth when the volcano is active. When Hephaestus placed the head of Typhon on his anvil, the monster's convulsions and attempts to escape produced earthquakes and volcanic eruptions. In another of his forges beneath the volcano Thera (Santorini), he manufactured armor for the gods and thunderbolts for his father Zeus—and the volcanic fires, smoke, flames, and shakings of the ground were the undeniable evidence of his industry.

This mythical view of volcanic eruptions was also expressed by the Greek poet Pindar in the fifth century B.C. After being defeated by Zeus in the war of the gods in the volcanic region known as the Phlegraean Fields in Italy, Typhon was by legend buried under Sicily:

> Now under sulphurous Cuma's sea-bound coast,
> And vast Sicilia's lies a shaggy breast;
> By snowy Aetna, nurse of endless frost,
> The pillar'd prop of heaven, forever press'd:
> Forth from whose nitrous caverns issuing rise
> Pure liquid fountains of tempestuous fire,
> And veil in ruddy mists of the noon-day skies,
> While wrapt in smoke the eddying flames aspire.

In Greek mythology, Hell was already considered hot, and located deep in the Earth. Plato (428–348 B.C.), in *Phaedo*, the deathbed drama of Socrates, gives a view of the geography of Hades, the netherworld of the Greeks and later the Romans: Hades is reached through a vast chasm that pierces through to the center of the Earth, where the high-walled prison Tartarus is found.[10] Inhabited by the dead—as well as a few gods and goddesses—the realm was governed by the god of the underworld, Pluto, and his queen Prosperina, a young woman he had seized near Mount Etna and carried down to reign with him in his fiery dominions.

In Roman times, the Sicilians still followed the ancient customs of ascending to the rim of Etna's crater to make offerings to the gods of the mountain. This practice, however, was condemned by Lucilius Junior, Emperor Nero's Procurator of Sicily. Protesting this ancient superstition, Lucilius very logically accounted for volcanic activity as a consequence of water seeping through fissures into the volcano where it came in contact with internal fires and was converted to steam—and was violently expelled during explosive eruptions.

The Hades of the Greeks and Romans made an easy transformation into Hell of the early

Christians—described in the Bible as a place with an eternal fire that will never be quenched (Mark 9:43–48). The Gospel of Matthew speaks of angels who throw the worthless "into the furnace of everlasting fire: there shall be wailing and gnashing of teeth" (Matthew 13:41,42). In the Book of Revelation, Hell contains both fire and brimstone (volcanic sulfur). At the end of the world, terrible earthquakes, volcanic eruptions, and scorched earth lead to the slaughter of two-thirds of all humanity. Finally, the Devil is flung into a lake of fire and sulfur. Lucifer the "firebearer" is linked to other fiery figures such as Prometheus and Icarus. After the fall, Lucifer becomes Satan.

In northern Europe, the Hell of the Vikings was called Niflheimur, a place of total darkness and cold. Like the Roman and Greek netherworld, it was also populated by giants, and in the southern part of Niflheimur was Muspell, a hot and fiery region ruled by the fire giant Surtur. In Icelandic mythology, Surtur may be considered the incarnation of volcanic eruptions. According to the tenth-century poem *Völuspá*, Surtur comes from the south with fire and fights the god Freyr at Ragnarök during the apocalyptic battle of the giants at the end of the world in Norse mythology. The climax of the battle, when "fire leaps high about Heaven itself," is a clear reference to the final destruction due to volcanic conflagration. The memory of Surtur is alive and well among the Icelanders; when a volcanic eruption broke out in the ocean south of Iceland in 1963, the new fiery island that rose out of the waves was named after the giant: Surtsey.

The association of mythical creatures with active volcanoes continues into the twentieth century. The Tolai people on New Britain island (Papua New Guinea) considered volcanoes inhabited by very powerful supernatural spirits called *Kaia*.[11] Volcanic eruptions were referred to as "a punuongo na Kaia," literally, the explosion of the spirit. Belief in the Kaia was still prevalent among the Tolai at the time of the 1937 eruption of Rabaul volcano, when over five hundred people were killed. The Australian volcanologist Wally Johnson tells of two elderly Tolai men who were not sure which supernatural being—God or Kaia—was responsible for the eruption: "Two old heathen sat in their hut in Vunapaka saying Our Father over and over again as well as they knew how. As the fall of pumice showed no sign of stopping, one said 'Turaqu, just go out and try your magic.' Turaqu crept out of the hut, took up his position, blew on his conch, and hurled his spells against the supposed Kaia. A flash of lightning passed in front of him, a violent thunderclap followed, and Turaqu vanished rapidly into the hut and said 'My friend, this hasn't been caused by Kaia, but by Him up above: we'd better go on praying'!"

Few countries have such an abundance of volcanic activity as Japan. In Japanese mythology the Oni monster is a horned, red giant who inhabits volcanoes and hurls out rocks during eruptions. He is still very much a part of popular Japanese culture in regions around active volcanoes, where a wealth of representations of Oni can be found in souvenir shops today (Figure 2.4). Many volcanic features in Japan are typically named *Jigoku*, or Hell. Jigoku is a term derived from Buddhism, and originally meant underground prison, but since medieval times it has come to signify the world of suffering, as opposed to the "paradise of supreme bliss"

2.4. The Oni monster of Japanese mythology inhabits active volcanoes and hurls out rocks during eruptions. This contemporary flag is sold in souvenir shops around Asama volcano, commemorating the great eruption of 1783. A hot pyroclastic flow from the eruption buried villages on the flanks of the volcano, killing 1151 people (author's collection).

where Buddha himself resides. The volcanic Jigoku of the Japanese is thus analogous to the hell or inferno of the Europeans.

The Hawaiian people have created a great wealth of nature-myths, which reflect primarily their close relationship with active volcanoes and the ocean. Their many volcano legends have been carefully documented by the American historian William Westervelt.[12] One of the earliest Hawaiian gods was Ai-laau, "the one who devours the forest." The Hawaiian people saw him constantly at work, as he laid waste the forests on the slopes of volcanoes, pouring out the hot and fiery lava from the bowels of the Earth. His principal residence was in the crater of Kilauea, but when the powerful fire goddess Pele arrived in Hawaii, Ai-laau trembled with fear and fled from the islands. Pele remains the current personification of volcanic activity in Hawaii, and whenever a volcanic eruption breaks out, the Hawaiians say that "Pele has come again." Pele first made fire-pits with the Po-oa, her magical spade, but most of them were near the ocean and of no use to her, as they quickly burst in great explosions of steam and sand, a reference to the explosions that result when hot lava flows into the ocean, causing the lava to disintegrate and producing great quantities of volcanic ash and sand. Finally Pele made herself a permanent fire-pit when she dug deep with the Pa-oa in Kilauea and established an everlasting supply where great quantities of lava continue to flow. When Pele was angry she stamped the floor of the fire-pit, causing earthquakes and eruptions (Figure 2.5).

Pele had a running battle with her sister, Na-maka-o-ka-hai, the goddess of the sea, an interesting mythological portrayal of the great clash of geologic forces between the Pacific Ocean and the Hawaiian volcanoes. In one of their battles, Na-maka-o-ka-hai rushed inland and tore the body of Pele apart, scattering her bones in great pieces along the coastline of the Haleakala volcano. Here the "bones" can be seen today as enormous blocks of columnar basalt. However, the spirit of Pele lived on in the great fire pits of Kilauea and Mauna Loa volcanoes and was worshiped by the Hawaiian people, who still bring her offerings of garlands, pigs, chicken, and fish. The goddess had a distinctly vengeful personality, however, and she was worshiped mainly by people whose lives were filled with a burning anger against their fellow humans. Characterized as an old and dangerous witch, she gave rise to the Hawaiian proverb "Never abuse an old woman; she might be Pele." When a part of a dead person was offered into Kilauea as a sacrifice, it was done in the hope that the spirit would become an *aumakua*, or a ghost-god with the power to aid vengeful worshipers in "praying" other people to death. The concept and name of Pele probably have an ancient Polynesian or East Indics origin. The fire goddess in Samoan culture bears the similar name Fee, and in Tahitian culture, Pere is the name for a volcano or a fire goddess. Further afield in Indonesia, the similar word *pelah* means hot and *belem* to burn.

2.5. An 1814 engraving of the crater of a Hawaiian volcano, as seen by the early European explorers: "Truncated mountain with the focus of an extinguished volcano, in the Island of Owyee." (author's collection)

No single event has boosted the Pele legend so much as the annihilation of the Hawaiian chieftain Keoua's army during the very rare and possibly unique explosive eruption of Kilauea in November 1790. War had broken out between Kamehameha, King of the island of Hawaii, and a group in the southwest part of the island, led by Keoua. During the skirmish, some of Keoua's army took a shortcut across the island and camped at dusk near the Kilauea pit-crater. During the night a series of great explosions sent up an enormous hot cloud of ash from the volcano. The violence continued, and Keoua was afraid to move his army through the danger zone; on the third day they cautiously began their march. No sooner had they started, when an even larger explosion took place, spreading a ground-hugging cloud of ash in all directions from the crater, enveloping them in total darkness within a burning hot and deadly surge cloud of ash.[13] The soldiers in the forward company were burned to death or asphyxiated by the cloud, while those in the rear were injured by falling stones and hot ash. Keoua survived, but his army was now in shambles, and the people of Hawaii regarded the event as proof that the fire goddess Pele had taken King Kamehameha under her special protection as the chief ruler of the island. Pele wreaked havoc once again during the eruption of Hualalai on the island of Hawaii in 1801, sending a great torrent of lava over villages and crop lands. Despite numerous offerings of pigs and produce thrown into the hot lava to appease the wrathful goddess, the lava continued on its destructive course. Finally, King Kamehameha went up to the flowing lava and flung into the fiery hot torrent the most precious gift he could offer—a large lock of his hair. Within a few days the lava had ceased its flow and the people attributed it to the great influence of their king over Pele.

3

The Bronze Age Eruption of Thera and Lost Atlantis

But afterwards there occurred violent earthquakes and floods; and in a single day and night of misfortune all your warlike men in a body sank into the earth, and the island of Atlantis in like manner disappeared in the depths of the sea.

Plato, ca. 350 B.C.

What a stroke of luck that the Mediterranean region, that cradle of Western civilization, is pockmarked with active volcanoes! There the early Greeks and Romans could observe local volcanoes and incorporate this phenomenon first into their mythology and later into their philosophy of nature. Their ideas regarding volcanic activity were born not in a vacuum or from secondhand accounts, but from firsthand experience.

Two major volcanic provinces in the Mediterranean region allow eruptions to be studied: the Hellenic arc, a line of islands stretching from Greece to Turkey that separates the Aegean Sea from the Mediterranean, and the Italian volcanic province, from Mount Etna in the south to Vesuvius in the north (Figure 3.1). Both regions owe their volcanic nature to plate collisions and result from the steady motion of the huge plates that make up the Earth's crust. As noted earlier, the African plate is drifting northward, gradually shrinking the Mediterranean basin and thrusting underneath the European and Asian plates. Where underthrusting occurs, in the regions known as subduction zones, conditions are ripe for the generation of magmas by melting in the deep Earth, a process that leads to volcanic activity at the surface. In the Hellenic arc the volcanic centers include, from west to east:

Aegina, Methana, Poros, Milos, Kimolos, Polyaigos, Pholegandros, Thera (Santorini), Nisyros, and Kos. Most have remained dormant in historical times, with the notable exception of Thera.

In Greek mythology, the volcanic island of Thera originated from a clod of earth. On their journey across Lake Tritonis the Argonauts encountered the local demigod of the sea, Triton, who presented them with a clod of earth. One of the Argonauts later dropped this gift into the sea north of Crete, where it formed the island; from its "origin" in myth, Thera or Santorini has remained the focal point of volcanism in the Hellenic arc for millennia. The most colossal eruption of Thera occurred in the seventeenth century B.C., when the Bronze Age Minoan civilization on Crete and in the Aegean Islands was at its height. This event—commonly referred to as the Minoan eruption—was the largest of its kind in the Mediterranean region since the great Campanian eruption in the Phlegrean Fields in Italy about 35,000 years ago. No contemporary accounts of the Minoan eruption exist, but much is known about the event from modern volcanological studies of the volcanic ash deposit, both on land and from cores drilled into the ocean floor sediments. The explosion sent into the atmosphere a 36-km-high (22 mi) volcanic plume of ash and pumice, most of which was transported dominantly to the east and southeast on the prevailing winds.[1] Today we find the Minoan ash layer in soils on islands as distant as Kos and Rhodes, on mainland Turkey, and detectable as far as the Nile delta in Egypt and even in the Black Sea.[2] On Thera, the excavations of Greek archaeologist Christos G. Doumas have revealed that the fallout layer of pumice and ash alone is about 7 m (23 ft) thick—enough to bury the flourishing Minoan

3.1. A map of the Mediterranean region, showing the location of the principal volcanic features that influenced the thinking of Greek and Roman scholars in antiquity about volcanic action.

town of Akrotiri.[3] The total volume of ash and pumice fallout from the eruption was equivalent to 20 cubic km (5 cubic mi) of rock, determined from the thickness and distribution of the layer on land, as well as in the sea, where the layer has been recovered from cores drilled into the sediment. Besides this fallout, vast quantities—another 20 cubic km (5 cubic mi)—of so-called hot ash flows or ignimbrites issued from the volcano and into the Aegean Sea, where they displaced a large mass of sea water, producing a huge wave, or tsunami, that likely inundated the coasts of nearby islands—perhaps devastating the densely populated north coast of Crete. Thus, we can estimate that the total production of magma during the eruption is about 40 cubic km (10 cubic mi) of rock, or 84 million tons, more than forty times the quantity produced by the well-known 1980 eruption of Mount Saint Helens.

It has also been proposed that the atmospheric effects of the Minoan eruption were truly global in extent. We know from observations of other large-scale explosive eruptions, such as the great eruption of Tambora volcano in Indonesia (1815), that volcanoes can cause short-term changes in global climate. Studies of growth rings, based on radiocarbon dating in ancient trees as far removed from the Mediterranean as California and Nevada and in Irish bogs, indicate a stunting of tree growth around 1626 to 1628 B.C., due to a sudden change in climate patterns that has been attributed to the Minoan eruption. More evidence of unusual conditions at this time comes from ice cores taken from the Greenland ice cap. The cores are recorders of annual snowfall on the ice cap during the past several thousand years, but the snow, now transformed into crystalline ice, also contains the fallout of impurities present in the atmosphere at the time, such as volcanic aerosols that are composed largely of sulfuric acid. A volcanic sulfuric acid layer occurs at a depth of 1227.5 m (4026 ft) in one of the south Greenland ice cores and has been dated to about 1645 B.C. We do not yet know if the layer in the ice dates from the Minoan eruption, but chemical analyses show that it represents global fallout of some 200 million tons of sulfuric acid aerosol droplets. During a large explosive eruption, volcanic gases such as sulfur dioxide may be transported to the Earth's stratosphere 20 to 40 km (12 to 24 mi) above the surface, where they react with water vapor and are transformed into tiny droplets of sulfuric acid. The fine mist of volcanic pollutants thus created acts as a veil against incoming solar rays, backscatters incoming solar radiation, and can alter the Earth's climate significantly for a few years.

The Minoan eruption of 40 cubic km (10 cubic mi) of magma drained a large subterranean reservoir under Thera, collapsing the island along circular faults and producing a caldera about 800 m (2600 ft) deep and 8 km (5 mi) in diameter. This massive cave-in reduced the volcano to a jagged circle of island fragments. In the 3500 years since that great eruption, there have been numerous small eruptions on the sea floor in the center of the submerged caldera, resulting eventually in the emergence of a new island in the second century B.C.

An eruption of the magnitude of the Minoan event must have created havoc with the environment throughout the eastern Mediterranean. The fallout, besides depositing a thick blanket of ash on islands to the east, probably destroyed or damaged crops. The tsunami caused by ash flows or pyroclastic flows previously mentioned must have had an even greater impact. When huge amounts of material issue from an island volcano into the sea, the great wave or series of waves thus created spread radially from the center of eruption. A modern example of this phenomenon was observed during the 1883 eruption of Krakatau in Indonesia, when pyroclastic flows with a volume equivalent to about 7 cubic km (2 cubic mi) of magma were nearly instantaneously discharged into the ocean. The tsunamis generated by that eruption were more than 35 m (115 ft) high as they swept the coasts of Java and Sumatra, about 50 km (31 mi) from the volcano, taking with them more than 36,000 lives. We know that Thera discharged a flow two to three times larger in volume, and it is reasonable to assume that the resulting wave was considerably larger than the Krakatau tsunami, even at a range of 100 km (62 mi), the distance of Crete from Thera. Another element has recently been identified as a major killer in the 1883 Krakatau eruption—awesome pyroclastic surges or dilute and very hot pyroclastic clouds that moved north from the volcano, across the sea, fatally burning 2000 people on the coast of Sumatra, 50 km (31 mi) from the volcano.[4] Did hot pyroclastic surges of this type also invade the north coast of Crete during the Minoan eruption? Study of both modern and ancient examples shows that while they are very destructive and lethal, pyroclastic surges may leave behind but little evidence of their passage.

At the peak of Minoan civilization, about 1600 B.C., the palaces on Crete and in major Minoan towns on other islands were suddenly destroyed, and the Minoans rapidly declined. What is amazing is that up to the moment of destruction, their civilization shows no sign of decline or impending collapse. It is therefore not surprising that archaeologists have long sought an external cause, such as a great natural disaster, to account for the ruin of these people.

Evidence of the Minoans was discovered almost single-handedly by the English archaeologist Sir Arthur Evans in the first years of the twentieth century. It was a discovery entirely inspired by the literal interpretation of Greek mythology. Evans was following in the footsteps of Heinrich Schliemann, the discoverer of the legendary city of Troy, who "set out with Homer in one hand and a spade in the other."[5] Literary evidence of the Minoan people on Crete and elsewhere in the Aegean abounds in the Greek classics, especially such epics as the *Iliad* and the *Odyssey*. In the *Odyssey* Homer states, "Among their cities is the great city of Knossos, where Minos reigned." In 1883 Schliemann obtained permission to dig at Kephala in Crete, suspecting it to be the site of the legendary Knossos, but excavations were delayed and had not begun by his death in 1890. In 1894 Evans acquired the Kephala site, and began excavations there in 1898, with immediate and spectacular success. He uncovered a huge building complex, clearly a major cultural and administrative center, that probably functioned more like the Kremlin than Buckingham Palace. In the process he discovered a completely unknown Bronze Age civilization that he labeled the Minoan, in keeping with his belief in Homeric legend, and historical substance was given to an area where legend had prevailed. On the basis of pottery styles in the excavations, Evans established a relative chronology of Lower, Middle, and Late Minoan for the sites on Crete and suggested that the Minoan period ended about 1100 B.C. When the eruption occurred, Thera was also populated by an advanced and wealthy Minoan culture, closely allied with the Minoans on Crete. They were seafaring people who maintained active trade throughout the eastern Mediterranean and had established prosperous communities centered on several palaces on Crete.

The wholesale and simultaneous destruction of these palaces greatly impressed Evans, who initially believed that the cause was a devastating earthquake, but later considered it a result of the Thera eruption when evidence of the volcanic event was discovered. It has been proposed by archaeologists that the destruction was mostly by fire, but at such sites as Amnisos the buildings show evidence of dislocation, which might be caused by a tsunami or an earthquake. Although some features of the destruction might be attributed to conquest by war and looting and sacking of palaces, most of the evidence suggests a natural disaster.

Earlier, in the 1870s and 1890s, French and German archaeologists discovered a large site, later known as Akrotiri, on the southern part of Thera, one that was clearly Minoan in age and buried by the great eruption. But it was not until 1939 that, impressed by the association of the volcanic deposit and abandonment of the site, the Greek archaeologist Spyridon Marinatos proposed that the sudden demise of the Minoan civilization on Crete was intimately connected with the eruption. Marinatos began excavations of Akrotiri in 1967, which have been continued to the present day with spectacular results.

Is it a mere sport of nature that one of the largest known natural disasters in the Mediterranean coincides approximately with the destruction and abandonment of the great palaces on Crete and the fall of the Minoan civilization? It is logical to propose a cause-and-effect relationship between the two major events, but if Marinatos' theory is to remain valid, it must pass a crucial test: a stratigraphic correlation between the destruction level on Crete and the level of the volcanic deposit at Akrotiri. This question is becoming one of the more exciting detective stories in modern archaeology. There are three basic approaches to the problem. One relies on the relative chronology of Minoan ceramic styles first developed by Arthur Evans. Just as in the modern world our tastes for pottery and manufacturing styles have changed over the centuries, so did the Minoan pottery styles evolve with time in a systematic manner. Another tool is radiocarbon dating of charcoal and other vegetation remains, and a third relies on finding the actual stratigraphic position of the volcanic ash fallout layer in the ruins of the palaces on Crete.

Ceramic chronology is the principal tool of the archaeologists, who rely on the rapidly changing fashions in ceramic styles. For example, what is known as the Late Minoan IB style was developed in a period of brilliant artistic and cultural achievements on Crete and one of its creative accomplishments was the distinctive and unique "Marine style" of pottery, often depicting life in the sea. Most of the evidence from pot shards seems to indicate that the destruction on Crete occurred during the Late Minoan IB period.

Yet on Thera the latest ceramics found in the town of Akrotiri date toward the end of the earlier Late Minoan IA ceramic phase, indicating that the eruption may be significantly earlier than the devastation on Crete.[6] Many questions remain to be answered, however, before we can equate the youngest pottery style found in Akrotiri with the actual time of the eruption. We know that ceramics were traded between Bronze Age cultures in the Aegean, between Crete and customers in Akrotiri. Could it be that pottery made in Crete reached more distant islands such as Thera many years after it came into general use in Crete? Is it also possible that Akrotiri was abandoned after such volcanic precursors as severe earthquakes long before the eruption, and no later pottery styles should be expected on Thera? How much time elapsed between that abandonment and the fall of volcanic ash and pumice? An analysis of soils found at Akrotiri, and underneath the pumice fallout layer, shows that the time between abandonment and eruption was sufficient for some soil formation, probably representing several years or even decades.[7] It is also known that the town had earlier been severely damaged by earthquakes and a lot of repair work was being carried out before it was abandoned. Major volcanic eruptions are generally preceded by severe but localized earthquake activity that may last for months or years. The Greek archaeologist Christos Doumas has suggested that one to five years of repair work took place after the earthquake and before final abandonment. No human or animal remains or skeletons of victims have been found beneath the volcanic deposit in Akrotiri, unlike in Pompeii and Herculaneum in Italy, where much of the population was interred by the volcanic deposit from the eruption of Vesuvius in A.D. 79. And it can be seen that the inhabitants of Akrotiri had removed most of their belongings, clearly indicating that the town was abandoned before the eruption. How long had it been empty? Several years before a great eruption, a volcano such as Thera will exhibit signs of activity, including frequent and even severe earthquakes, gas emissions, and small explosions. Akrotiri is virtually a stone's throw from the edge of the great caldera of Thera and was abandoned after such early unrest in the volcano. Allowing for a significant period for final desertion of the site, the ceramic chronology does not rule out a correlation between the eruption and the destruction on Crete.

Dating of the eruption and of destruction levels in Crete is in principle made possible by the radiocarbon method, but results have been somewhat disappointing. Part of the problem is that the Minoan volcanic deposits were not sufficiently hot to carbonize all vegetation material, and consequently dating errors are large. Ninety-four radiocarbon dates of seeds and other plant remains found in jars at the base of the volcanic deposit at Akrotiri give an extensive range, from 1750 to 1550 B.C. A critical assessment of this evidence favors a seventeenth-century B.C. date for the eruption and for the Late Minoan IA ceramic phase on Thera, but a lower, mid-sixteenth century B.C. age cannot be ruled out.[8] The Late Minoan IA or B periods have not been dated well on Crete, but the subsequent Late Minoan II period is now well radiocarbon-dated in Knossos in Crete at around 1510 to 1430 B.C.[9] Thus, the radiocarbon method has not yet resolved the question whether the eruption and the destruction on Crete are simultaneous.

In 1975 I participated in a study of ocean floor sediments in the eastern Mediterranean to determine the thickness and extent of the Minoan volcanic ash layer in the region as a whole.[10] The sediment cores taken during our cruise showed that the ash fallout should have been at least 5 to 10 cm (2 to 4 in) thick on Crete following the eruption. On the nearby island of Rhodes, the tephra layer is 20 cm (8 in) thick and falls at the boundary between Lower Minoan IA and IB; there is also evidence of abandonment there.[11] On Crete, however, only traces of ash had been found—it seemed that wind and rain erosion had removed the volcanic fallout almost everywhere. In the last decade, however, evidence of the Thera deposit has been found in at least seven sites on Crete, and this discovery has the potential for resolving the much-debated issue of the relationship between the eruption and the destruction of the Cretan palaces.[12] At Molchos in Crete the ash layer is clearly beneath the Lower Minoan IB pottery level, strong evidence that the eruption predated the destruction on Crete.[13] Given the present state of knowledge, it appears that the occurrence of the eruption, within years or decades of the destruction of the palaces on Crete, is a mere coincidence. One could, of course, postulate that the effects of the volcanic event weakened the Minoan kingdoms and that invading armies overcame them a few years later, but this will remain in the realm of speculation.[14]

The magnitude of the Minoan eruption was such that it must have been noted in Egypt and other countries around the Eastern Mediterranean. We know from the distribution of the ash fallout that the dispersal of ash was to the east and southeast, including the Nile delta. The eruption occurred during the Eighteenth Dynasty of Egypt, and several inscriptions on papyri contain descriptions that may refer to the eruption and its devastating effects. The Ipuwer papyrus in the Museum of Leiden contains these remarks: "Plague is throughout the land. Blood is everywhere. The river is blood. Men shrink from tasting and thirst after water. All is ruin! Forsooth gates, columns and walls are consumed by fire. The towns are destroyed. Oh that the Earth would cease from noise, and tumult be no more. Trees are destroyed. No fruits or herbs are found....The land is without light." The Hermitage papyrus preserved in St. Petersburg contains this passage: "The sun is veiled and shines not in the sight of man. None can live when the sun is veiled by clouds. None knoweth that midday is here...his shadow is not discerned. The sun is dim in the sky like the moon. The river is dry, even the river of Egypt. The Earth is fallen into misery. This land shall be in perturbation." An inscription in the shrine at El Arish reads: "The land was in great affliction. Evil fell on this Earth.... It was a great upheaval of the residence. Nobody left the place during nine days, and during these days of upheaval there was such a tempest that neither the men nor the gods could see the faces of their next."

These fragmentary Egyptian writings are similar to the biblical descriptions of the plagues of Egypt, and indeed J. G. Bennett has suggested that these plagues, which paved the way for the Exodus, were actually the environmental effects of the Minoan eruption.[15] During the New Kingdom a group of Semitic people, called Hebrews by the Egyptians, followed their leader Moses out of Egypt and into the deserts of Sinai. The date of the Exodus is not known, but may have been around 1500 B.C., roughly contemporaneous with the Thera eruption.[16] Exodus describes one of the plagues as follows: "And all the waters that were in the river turned to blood. And the fish that was in the river died, and the river stank, and the

Egyptians could not drink of the water of the river, and there was blood throughout all the land of Egypt." (Exodus 7:20–22) The fall of pinkish-grey ash from the Thera eruption over Egypt could account in part for these remarks. Exodus remarks again: "And there was thick darkness in all the land of Egypt for three days. They saw not one another, neither rose any from his place for three days." (Exodus 10:22–23)

Although no written records exist of the Minoan eruption, it is possible that an account of its destruction is preserved in two Greek legends: that of the lost continent of Atlantis and the battle between the Olympian gods and the Titans, as recorded in the poem *Theogony* (Descent of the Gods) by the Greek poet Hesiod (ca. 700 B.C.). About 350 B.C. Plato documented in *Timaeus* the destruction of Atlantis, citing the Athenian statesman Solon as a source, who in turn had heard the tale from an Egyptian source in about 600 B.C. Plato's Atlantis was a great and prosperous empire that came to an end because of a natural catastrophe. He wrote, "But afterwards there occurred violent earthquakes and floods; and in a single day and night of misfortune all your warlike men in a body sank into the earth, and the island of Atlantis in a like manner disappeared in the depths of the sea. For which reason the sea in those parts is impassable and impenetrable, because there is a shoal of mud in the way; and this was caused by the subsidence of the island." It is said that at the heart of every legend there lies a seed of truth, and it is in this spirit that countless books have been written about Atlantis. In 1909 the Irish scholar K. T. Frost, in an article called "The Lost Continent," proposed that lost Atlantis was the Minoan civilization on Crete.[17] He was the first to point out the striking similarities in the culture of Atlantis as described by Plato and the Minoans on Crete, but his hypothesis lacked a mechanism to account for the disaster. This was later provided by the volcanic hypothesis of Marinatos in 1939, and this version of the Atlantis legend has been the subject of several books.[18]

In the *Theogony* Hesiod presents the mythology and genealogy of the Greek gods, which equates to the cosmology and creation of the world of the ancient Greeks. Although Hesiod was born on the Greek mainland, his father originated from Cymae on the Ionian coast, now southwestern Turkey.[19] Hesiod's family thus had a geographic experience that spans the active Hellenic volcanic arc, including the island of Thera. Hesiod's poem is the earliest Greek literature devoted to a mythological topic, but for us it acquires a special significance in the recent interpretation of the American historian of science, Mott Greene, who has proposed that the great battle between the Olympian gods and the Titans represents a volcanic eruption, specifically the explosive eruption of Thera.[20] As mentioned previously, Greene convincingly argues that the sequence of events in the battle and its effects on the physical world can be most logically interpreted as the recounting of the catastrophe.

In the battle the Olympian gods are led by Zeus, and the leader of the Titans is Kronos (Figure 3.2). A great number of natural phenomena are mentioned, which could be interpreted as the effects of volcanic eruption, and their temporal sequence is consistent with this hypothesis. The poem's description of a long war preceding a climactic battle represents pre-eruption seismic activity in the volcano. Increasing seismic activity is indicated in the poem

3.2. The earliest surviving fragment of Greek literature is the sixth-century B.C. poem *Theo-gony* by Hesiod. It describes the battle between Zeus and the Titans, which has been interpreted to portray the great eruption of the volcano Thera that occurred about 1000 years before the poem was written. In this etching of the "Fall of the Giants" (1663) by the Neapolitan artist Salvator Rosa, the giants and rocks tumble down from Olympus, with Zeus flying above in the clouds. The etching is inscribed: "*Tollontur in altum, ut lapsu graviore ruant*" (They are raised up high, that they may be hurled down in more terrible ruin). Etching from the Rijksprentenkabinet, Amsterdam.

by both sides gathering strength. "The limitless expanse of the sea echoed terribly" would signify the first phase of the eruption and "the broad area of the sky shook and groaned" could be interpreted as atmospheric shock waves from the first explosion. "Missiles thrown" and "weapons discharged at each other whistled through the air" signify fragments of volcanic rocks, or pyroclastic ejecta, and volcanic lightning heralds the arrival of Zeus on the battlefield. Finally, when "the whole earth and the ocean streams and the barren sea began to boil," the poet referred to the passage of hot pyroclastic flows over the ocean's surface, up to a hundred kilometers from the volcano.

The poet Robert Graves was actually the first scholar to propose a connection between the Thera eruption and the legend of the battle of the Greek gods and the giants.[21] Greek legend tells us that when the giants came rushing toward Olympus the gods fled in terror to Egypt. Graves suggested that the event may have marked the exodus of frightened priests and priestesses from the Aegean archipelago as a result of the devastating eruption at Thera. A description of the fall of the Titans is also found in the *Metamorphoses* of the Roman poet Ovid (43 B.C.–A.D. 17):

> And, that high heaven might be no safer than the earth,
> They say that the Giants essayed the very throne of heaven,
> Piling huge mountains, one on another, clear up to the stars.
> Then the Almighty Father hurled his thunderbolts,
> Shattered Olympus, and dashed Pelion down from underlying Ossa.
> When those dread bodies lay o'erwhelmed by their own bulk,
> They say that Mother Earth, drenched with their streaming blood,
> Informed that warm gore anew with life, and,
> That some trace of her former offsping might remain,
> She gave it human form.

The eruption of Thera did not occur in isolation, however, because in Italy Vesuvius erupted at about the same time. Radiocarbon dates of the great pumice deposit found around Vesuvius, from an explosive event known as the Avellino eruption, give an age date that is indistinguishable from the Thera event.

Thera volcano has always been one of the most active volcanoes of the Mediterranean and since the Minoan event, many eruptions on the submarine floor of the caldera have resulted in the emergence of several new islands. Strabo reported in the *Geographia* (Book I.3.16) that in 197 B.C., "Midway between Thera and Therasia fires broke forth from the sea and continued for four days, so that the whole sea boiled and blazed, and the fires cast up an island which was gradually elevated as though by levers and consisted of burning masses—an island with a stretch of twelve stadia [2 km; 1.6 mi] in circumference. After the fires ceased, the Rhodians who then controlled the seas, first dared to approach the place and to erect on the island a temple to Poseidon Asphalios." The new island was named Hiera (the sacred one) by contemporary Greeks, but is now known as Palea Kameni (the old burnt one).

An eruption near Thera is mentioned in a passage in *The Life of Appolonius* (175, IV, 34) by the Greek philosopher Philostratus. The mystic Apollonius, journeying in Crete, was paying a visit to the temple of Asclepius at Lebena near Phaistos when an earthquake struck. A thunderous noise came from the ground and the sea retreated nearly a mile from the shore. Apollonius observed this and cried out: "The sea has given birth to land." A few days later the news came that on the day of the portent, a volcanic island was cast up in the strait between Thera and Crete. In A.D. 19 another eruption in the sea produced the rocky islet of Theia (divine island). Some sources refer to an eruption in A.D. 46 that also produced a small island. Theophanes describes an eruption in A.D. 726 that created yet another new island, one that coalesced with Hiera. The volcano showered the town of Thera with pumice, and ash was carried by the winds to the coast of Turkey.

In 1570 the island of Mikra Kameni (small burnt one) formed during a long eruption, and just a century later (1650) a large eruption took place near Kolumbos, outside the Thera caldera, in the sea just northeast of the main volcano. The fireworks from this event were visible from Crete and explosions were heard in the Dardanelles, 500 km (310 mi) away, tsunami were generated, and pumice floated to distant islands. The eruption emitted huge quantities of noxious gases, which caused the death of a thousand livestock and many people on Thera. Later, from 1707 to 1711 eruptions beneath sea level inside the caldera built up Nea Kameni (new island) between Palea Kameni and Mikra Kameni. In June 1707 a whitish rock emerged from the sea within the caldera, draped with wet sediment. When the local inhabitants landed on the rock they found a multitude of full-grown live oysters adhering to it, which they promptly ate. The shells brought up during this upheaval of the sea floor normally live at depths in excess of 67 m (220 ft), indicating the minimum amount of the rapid uplift that must have occurred. This uplift was soon followed by lava erupting from the new island, with minor explosive activity. Further eruptions occurred in 1866–70, 1925–26, 1928, 1939–41, and the most recent one in 1950. During these eruptions, the former tiny islands have been united with the larger island of Nea Kameni, which stands 130 m (426 ft) above sea level.

The Chariot of the Gods

One of the earliest known observations of a volcanic eruption comes from the voyage of the Phoenician explorer Hanno the Navigator to West Africa. The Phoenicians, a Semitic people living on the coast of Lebanon, were seafarers and traders, with colonies throughout the Mediterranean. They may even have ventured as far as Cornwall in England in search of tin. They flourished after the decline of the Minoan empire, building important cities around the Mediterranean such as Byblos, Tyre, and Sidon.

In about 430 B.C., when the Phoenician empire was at its peak in the city of Carthage (Tripoli), Hanno led an expedition of sixty vessels from that city through the Pillars of Hercules (Straits of Gibraltar) and south around the coast of western Africa, with the aim of founding new colonies.[22] Hanno commemorated his voyage with an inscription engraved on a stele in the Baal Hammon temple of Cronos at Carthage (the name in Phoenician means new town). The inscription was translated into Greek, probably in the fourth century B.C. and the details of the fantastic expedition have been preserved in Hanno's account, known as the *Periplus*.[23] As Hanno and his men sailed along the coast, they came to a shore where streams of liquid fire flowed into the sea. The land was inaccessible because of the intense heat, and the Carthagians sailed on, but for four nights they could see flames.[24] "We saw at night a land full of fire. In the middle was a lofty fire larger than all the rest touching seemingly the stars. By day this appeared to be a very great mountain called Chariot of the Gods." The only volcano that could fit the location and this description is Mount Cameroon, a very large (4095 m; 13,432 ft) and frequently active volcano that juts out from western Africa into the South Atlantic Ocean on the border between Nigeria and Cameroon. In the seventeenth century, the English scientist Robert Hooke took an interest in this vivid account and proposed that Hanno had in fact witnessed the final stages in the disappearance of Atlantis.[25]

Mount Cameroon is by far the most prominent of a string of volcanoes, called the Cameroon line, that extend from the Atlantic Ocean deep into the African continent, terminating in the Saharan desert to the northeast. Among its many eruptions in historical times were the events in 1909, 1922, and 1959. The 1922 eruption was particularly destructive, as the lava flows burned down large forests and destroyed many farms before reaching the ocean. The Bakweri people who inhabit the slopes of the volcano had always viewed it with a mixture of awe and terror, even resorting to human sacrifice. During eruptions as late as the early part of the twentieth century it is reported that they attempted to appease the angry spirit of the volcano by strapping the albinos of their tribe to stakes driven into the ground in the path of the advancing fiery lava flows. The latest eruption of Mount Cameroon, in 1982, produced a stream of lava that descended the south flank of the volcano and flowed for 12 km (7.5 mi) toward the ocean. The Bakumeii people call the volcano *Manga-ma-loba* (Cavern of the Gods), and make much of the fact that three of its four eruptions this century have coincided with the death of a tribal leader.

The Cameroon line includes two lake-filled craters that were the scenes of mysterious events in 1984 and 1986 that took the lives of thousands of the local people. In 1984 I received a telephone call from the U.S. State Department requesting that I travel immediately to West Africa to investigate the sudden and mysterious death of thirty-seven people around Lake Monoun on the Cameroon volcanic line. (Earlier in 1979 I had acted as a consultant for the State Department at the volcanic eruption of Mount Soufriere on the island of St. Vincent in the Caribbean.) When my graduate student Joe Devine and I reached Cameroon we found that the people had died in their tracks as they walked along the shore of the Lake Monoun crater in the predawn hours of August 15, 1984, on their way to market in the nearby village of Foumbot. Their bodies were found later that morning, lying by the roadside, totally unmarked. Various rumors about the cause of their deaths circulated among the U.S. Embassy staff, Cameroon government officials, and local scientists. Speculations included poisoning of the people by a local politician, a manmade explosion in the lake, and a volcanic explosion.

When we began our study of the lake, we found that it was indeed a very deep and bowl-shaped volcanic crater basin, with a depth of 90 m (295 ft). As we started to haul water samples from the bottom to the surface and raise them into our dinghy, we discovered to our dismay that the deep waters were extremely gas-rich and discharged a great deal of an odorless and colorless gas from the sampling bottles. Yet we found no sign of heat in those waters and therefore no indication of a volcanic eruption at the time of the deaths. In the light of the new evidence, I began to develop another hypothesis involving a high carbon dioxide content in the lake, which was later supported by chemical tests carried out in our laboratory in the United States. If the deep waters of Lake Monoun were saturated with carbon dioxide before the disaster, then a sudden overturn, or inversion, of the lake water would trigger a catastrophic release of gas due to the decrease of pressure. During an overturn, deep, gas-rich waters would be brought suddenly to the surface, experiencing a great drop in pressure. This leads to explosive exsolution of the carbon-dioxide gas from the deep waters to the atmosphere. The gas would spread as a thick, coloress, and odorless cloud around the lake, and then pond into depressions because it is somewhat more dense than the atmosphere. This gas burst must have occurred during the night and when the unsuspecting villagers made their way with their wares to market in the predawn hours, they walked directly into a low depression by the lakeshore and were quickly suffocated in the insidious oxygen-free carbon-dioxide cloud. We proposed that, over time, carbon-dioxide gas leaks out of the volcano onto the lake floor, but is dissolved in the deep water because of the high pressure. Just as pressure holds carbon dioxide in solution inside an unopened bottle of champagne or a can of soda, so does the pressure exerted by water on the bottom of the lake result in a very high content of carbon dioxide dissolved in the deep water. The real puzzle was the process that brought about the "overturn" of the lake waters. We suggested several mechanisms, with the most likely explanation that a landslide had plunged into the lake, pushing deep water upward and leading to spontaneus exsolution of gas that was already at saturation. This created a zone of bubbly and therefore low-density water in the lake, which in turn led to a decrease in the water pressure, causing a runaway process, or chain reaction, of degassing from other parts of the lake waters saturated with carbon dioxide.

Although all the evidence supported this idea, the carbon dioxide theory seemed far-fetched, arousing skepticism that such a process could occur. After lengthy delays by the reviewers, a paper I wrote proposing the idea was finally accepted for publication,[26] but just as it was appearing in print the mysterious Cameroon line struck again! In 1986 an even larger gas burst issued from Lake Nyos a few kilometers distant from Lake Monoun. Nyos is larger and over 200 m in depth, and the resulting carbon dioxide gas cloud was of much greater volume. The invisible and tasteless cloud flowed into a densely populated valley during the night, enveloping several villages and suffocating 1746 people. All the evidence from Lake Nyos suggested the same answers as in the case of Lake Monoun two years earlier, but we are still struggling with the fact that two such unique lethal events occurred only two years apart; was it a mere coincidence or is there a common underlying cause that we do not yet comprehend?

4

Subterranean Wind and Internal Combustion

What takes place in the Lipari Isles affords an additional proof that the winds circulate underneath the Earth.

Aristotle

From the dawn of history, Mediterranean people have witnessed eruptions of volcanoes. Because of its unique geologic structure and high level of activity, this region had already fostered an interest in volcanic phenomena and other Earth processes in antiquity. As noted earlier, the thick continental African plate pushes inexorably northward at a rate of about 1 cm (0.4 in) per year, and the weak floor of the Mediterranean Sea is deformed, folded, and thrust down into the depths of the Earth's mantle. This has led to a string of volcanoes along the zone of weakness, forming two major volcanic subduction zones: the Aeolian island arc in the west and the Hellenic island arc in the east. The structure of these volcanic arcs is highly complex because of the great deformation of the Mediterranean crust and the adjacent continental plate margins. However, because of the central role these volcanoes play in the evolution of ideas about volcanic processes, it is appropriate to examine their structural setting and eruptive history.

The active Aeolian Islands, Stromboli, Lipari, and Vulcano, belong to an arc of volcanoes that has grown on top of the junction between the African and Eurasian plates in the southern Tyrrhenian Sea. All of these volcanoes have been active since antiquity and Stromboli has been continuously active as far back as recorded history, more than 2500 years (Figure 4.1). The Greeks first began to settle the Aeolian Islands and Sicily in the eighth century B.C. and had ample opportunity to observe volcanic activity. The "wind" or uprushing of air and gases during explosive volcanic eruptions in these islands so impressed early Greek seafarers that they considered them the home of Aeolus, a giant imprisoned in a cave beneath the islands and known as the keeper of the winds. The association of winds and volcanoes was to become a fundamental aspect of Greek ideas regarding volcanic activity.

By the fifth century B.C. the Greek philosophers had moved beyond the mythological view of natural phenomena, but their explanations of volcanic activity were highly speculative and, unfortunately, only fragments of their writings on this topic remain. Their theories of volcanism are intimately linked to the concept of the origin of heat in the Earth. Among the Greeks, Thales of Miletus (624–546 B.C.) has been called the first philosopher and the first scientist. None of his writing has survived, but we know that he developed the idea that the material world is basically intelligible and that the cosmos is an entity the workings of which the human mind will one day understand. He showed that the power of the human mind is reason, and it was his rationality that laid the foundation of "thinking about the world the Greek way." Some of Thales's ideas regarding the philosophy of nature have been pre-

4.1. Stromboli volcano in the Aeolian Islands of the Tyrrhenian Sea has been continuously active for more than 2500 years. Its activity is so persistent that sailors have referred to it as "the lighthouse of the Mediterranean." In ancient times it was known as Strongyle, "the round one," in reference to its near-perfect conical shape. This engraving, of its 1874 activity, is reproduced from J. W. Judd, *Volcanoes* (New York: D. Appleton & Co., 1881).

served in later works. He considered the Earth as resting on water and floating on its surface, rather like a ship moving on the sea. Seneca quoted Thales as follows: "It is because of the movement of the water that the Earth moves in what is called an earthquake." We do not know what Thales may have written on volcanoes, but later ideas about volcanism were inevitably linked to the concept of fire. The first philosopher to discuss the concept of fire and give it a prominent position in the system of natural philosophy was Heraclitus (546–484 B.C.) from the Greek colony Ephesus in Turkey. He saw fire as primal matter that is transformed in infinite situations—the active agent that produces change in so many technical and natural processes. The order of the world, he proposed, was laid down by neither god nor humans, but ever was and ever shall be an eternal living fire, ignited and extinguished according to certain measures.

Perhaps the earliest ideas on volcanism were put forward by Pythagoras of Samos (ca. 540–510 B.C.), who founded a school at Croton in Italy. He believed in a central fire within the Earth, one that would eventually exhaust its fuel and die out. Another early scholar who pondered volcanic eruptions was the Greek Anaxagoras of Clazomenae (ca. 500–428 B.C.), who said that a mysterious substance called *ether* sank into the hollow interior of the Earth, where it mixed with vapors, causing lightning and fire, which were forced toward the surface.[1] Anaxagoras also claimed that the natural motion of air within the Earth was upward, but when its passage was hindered by some blockage of the porous parts, it was forcefully expelled from the Earth, causing earthquakes, tremors, and volcanic eruptions. His ideas on volcanism as an expression of "wind" within the Earth were the cornerstone of the interpretation of eruptions until Roman times. Later, Democritus of Abdera (ca. 400 B.C.), known as the founder of atomic theory, also regarded trapped air as a possible agent, but attributed earthquakes and volcanic eruptions primarily to the action of water moving at high velocity through narrow passages within the Earth.[2] His contemporary Archelaus (d. 428 B.C.) followed Anaxagoras, proposing that air is forced into the Earth through caverns and underground passages, where it is compressed to cause violent internal storms. When the compressed air escaped to the surface, it resulted in earthquakes and violent disturbance.

The Greek philosopher Empedocles of Agrigentum (ca. 495–435 B.C.) lived in Sicily, where Mount Etna in action was a common sight. He maintained that by uniting the four elements of fire, air, water, and earth in different proportions, he could explain all the endless varieties of substances known to humans. Empedocles' ideas about volcanism were quite modern. The Earth, he believed, had a molten center, and volcanic eruptions resulted from the rise of the molten material to the surface. Plato conceived of a subterranean river of fire, Pyriphlegathon, as the source of lava streams, and he followed Democritus in proposing that the driving force of the eruptions was air trapped under great pressure within the Earth.[3]

Aristotle (384–322 B.C.) is regarded by many as the father of natural history, and his far-ranging contributions to the study of the Earth also included some sagacious ideas on volcanic activity. These he presented in Chapters 7 and 8 in Book II of his great work *Meteorologica*, a broad discussion of the physical universe. In particular, he discussed the origin of earthquakes, attributing the same or similar origin for volcanic eruptions.[4] But in order to comprehend Aristotle's ideas of volcanic activity, we must look at his view of the physical world. All matter, he wrote, is composed of mixtures or combinations of the four elements, earth, water, air, and fire. He identified a fifth element, quintessence, which made up the celestial bodies, such as the fixed stars, the Sun, and the planets beyond the Moon. He proposed that heavy matter, composed mainly of the the earth element, tends to seek a place downward, toward the center of the Earth, while light substances, composed mainly of fire and air, tend to "fly" upward, above the Earth. Because air can be mixed in various proportions with both earth and fire, its behavior is unpredictable and highly perturbed. In a pure world where there is no wind, earth would be at the base, with water above it, overlain by air, and fire above all. "The Earth," he said, possesses "its own internal fire" that generates wind inside the Earth by acting on trapped air and moisture, leading to earthquakes and volcanic eruptions. He even makes a comparison with human anatomy in discussing the effect of the "internal wind": "For we must suppose that the wind in the earth has effects similar to those of the wind in our bodies whose force when it is pent up inside us can cause tremors and throbbings." Aristotle did not, however, propose a fire raging in the Earth's interior, but thought that the heat associated with volcanic eruptions was generated by the friction produced when the wind moved rapidly through restrictions within the Earth. "The fire within the Earth can only be due to the air becoming inflamed by the shock, when it is violently separated into the minutest fragments. What takes place in the Lipari Isles affords an additional proof that the winds circulate underneath the Earth."

Aristotle describes the Earth's crust at Heraclea in Pontus and an event at Hiera (Vulcano) in the Aeolian Islands of Italy as swelling up into a mound that finally bursts asunder by "winds" escaping from the Earth's interior. Aristotle refers here to the eruption of Vulcano in 360 B.C.: "For earthquakes have occurred in some places which have not let up until the wind which caused them had passed like a gale into the atmosphere through an eruption. Indeed this is what occurred near Hiera, one of the Aeolian Islands. For there part of the earth swelled up and, making a sound, it formed a kind of mound which finally broke open and let out a great wind which was laden with fiery cinders. The city of the Lipareans which was not far distant was reduced to ashes." Aristotle clearly regarded volcanic eruption as a continuum with earthquake, and he applies the term earthquake to such eruption. Thus, he recognized one type of earthquake in which "the shock runs from below, like a throb. Whenever this type of earthquake does occur, large quantities of stones come to the surface, like the chaff in a winnowing sieve. This kind of earthquake it was that devastated the country around Sipylos, the so-called Phlegraean plain and the districts of Liguria."

The identification of volcanic activity with "wind and fire" by the early Greeks follows logically from actual observations of a volcanic eruption. The most striking phenomena are first the explosive uprush of hot gases or "wind" from the Earth through the crater; then the incandescent red-hot glow of molten rock, giving the appearance of fire. On the basis of these phenomena Aristotle postulated a vast store of pent-up wind within the Earth, which generated friction and high heat and found its escape in volcanoes. After all, the Platonic view was that heat is a kind of motion. Aristotle's concept of volcanism is thus firmly rooted in his ideas about the capacity of motion to generate heat: "We see that motion can rarefy and inflame air, so that, for example, objects in motion are often found to melt." This theory, which dates to his predecessors Anaxagoras, Plato, and Democritus, was remarkably long-lived and had adherents well into the sixteenth century.

In 282 B.C. the volcano Methana erupted in the Sardonic Gulf. Because of its proximity to the city of Athens, this eruption must have greatly influenced the classical Greek contemporary philosophers and other students of nature. Strabo described the eruption: "And near Methana in the Hermionic Gulf the land was upheaved to a height of seven stadia as the result of a fiery explosion. This area was inaccessible by day because of the heat and the sulfurous odor, while at night it had a fragrant smell and glowed from a great distance. It was so hot that the sea seethed for five stadia and was turbulent for up to twenty stadia, and there appeared a heap of huge broken rocks, their size not less than towers."

The northern most of the Aeolian Islands, Stromboli, is the most isolated volcano in the island arc (Figure 4.2). The cone-shaped Stromboli was also known as *Strongyle* (the round one) in ancient times. Early references to its activity were made by Callius in the third century B.C., and Polibius mentions Stromboli as being active in his time. Nearly constant activity was reported by Cornelius Severus, Strabo, and Elder Pliny. South of Stromboli is the large island of Lipari, which owed its prosperity up to the Bronze Age to a great and inexhaustible

4.2. An 1814 engraving from the volcanic island of Stromboli in the Tyrrhenian Sea. (author's collection).

store of the brittle and hard volcanic glass obsidian. Most of the obsidian artifacts in the Mediterranean region during the Neolithic Age (6000 to 7000 years ago) are derived from the Pomiciasso obsidian lava flow on Lipari, which erupted about 8600 years ago. Lipari last erupted about A.D. 580, when explosions in the craters of Monte Pilato and Forgia Vecchia on the northwest coast spread a thick blanket of light grey ash over the island, followed by extrusions of obsidian lavas. These eruptions are documented in written record and legend, and the two craters were still active when San Willibald visited Lipari in A.D. 729. Legend has it that a pious hermit, Saint Calogero (A.D. 524–562), exorcised devils and their fire from the black obsidian stone of Lipari, driving them to the nearby island of Vulcanello, then to the Fossa crater of neighboring Vulcano—essentially the sequence of migration of volcanic activity in historic time.

The earliest historic eruption of Vulcano, in 424 B.C., was reported by the historian Thucydides, who was based in Sicily from 424 to 403 B.C. and described the explosive activity and its ejection of incandescent material. As noted earlier, Aristotle reported the next eruption in Vulcano in 360 B.C., and Theophrastus states that rumblings from the violent explosive eruption were even heard in Taormina on Sicily. Vulcanello was a small volcanic island situated between Lipari and Vulcano, but became enlarged and joined to Vulcano through sixteenth century volcanic activity. Livy described its formation in 183 B.C. when "not far from Sicily a new island issued from the sea which had not before existed." The fifth-century historian Orosius also mentions the eruption: "Then in Sicily the island of Vulcano which had not existed before, suddenly appeared from the sea amid great wonder and remains until now." And, writing in the first century, Elder Pliny says of the eruption: "Before our time and near Italy, there emerged from the sea an island among the Aeolians."

On Sicily, just south of the Aeolian Islands, towers Mount Etna, at 3000 m (9840 ft) the highest volcano in Europe and one of the largest volcanoes on Earth. Rarely dormant, it has been almost continuously active for over the 2500-year period that historical records cover. Its summit cone, which consists of interbedded lavas and scoria layers, has a basal diameter of about 2 km (1.2 mi) and is about 260 m (850 ft) high. The central crater at the top is 500 m (1640 ft) in diameter. Superficially, Etna seems to belong to the Aeolian volcanic arc, but more likely is of the class of large volcanoes called "hotspots" that derive magmas from the deep mantle of the Earth, independent of the motion of large crustal plates. The volcano produces two types of eruptions: summit crater eruptions, usually of small lavas and pyroclastics, and eruptions on the volcano's flank, with a much higher volume and effusion rate of lavas. Flank eruptions are of much greater concern for people living on the slopes and lowlands around the volcano because they destroy agricultural land, towns, and roads. Most such eruptions have issued from Etna's rift zones, which are systems of deep fissures that split the northeast, west, and south-southeast flanks of the volcano. About twenty eruptions of Etna were known in antiquity, but historical accounts date to about 1000 B.C., when the Phoenicians first established trading posts and settlements in Sicily. The name Etna may indeed be derived from the ancient Phoenician word *athana*, furnace.

In 734 B.C. the Greeks arrived at Naxos at the foot of Etna, rapidly colonized the east coast of Sicily, and soon took notice of Mount Etna. The volcano quickly entered into Greek legend, with the convulsions of the mountain seen as the violent struggles of the giant Typhon (Enceladus), attempting to throw off the burden hurled at him by Zeus after the war between the gods and the Titans (Figure 2.3). The Roman poet Virgil (70–19 B.C.) described it in his epic poem the *Aeneid* :

> T'is said Enceladus' huge frame,
> Heart-stricken by the avenging flame,
> Is prisoned here, and underneath
> Gasps through each vent his sulphurous breath:
> And still as his tired side shifts round,
> Trinacria echoes to the sound
> Through all its length, while clouds of smoke
> The living soul of ether choke.

The *Aeneid* was to have an enormous influence on general ideas of later writers about the underworld, including Dante and Saint Augustine. The poem is a description of the land of the dead, and was written between 30 and 19 B.C. when the poet was living in Rome. The poem's hero is Aeneas, whose long and perilous wanderings are described in twelve books designed to celebrate the Roman Empire. In Book 6 of the epic, Aeneas lands at Cumae in Italy, where he visits the famed priestess Sybil, who resides in a cave at Avernus near Cumae volcano in the Phlegraean Fields (Greek, *fleguros* or burning —a region of numerous eruptions, steaming hot springs, and sulfurous fumaroles).[5] Together they descend through this cave, crossing the fiery River Styx, into the underworld. There they encounter various groups of the dead, reach the gates of Tartarus where the worst sinners are punished, and finally depart through the Ivory Gate—by which false visions are sent to mortals.

Ever since, the volcanic regions of the Phlegraean Fields and Cumae have represented a fundamental point in the search for the underworld. The myth that Virgil so poetically narrated dated to long before Roman times and had its roots in the volcanic landscape of the area. Centuries earlier, Homer wrote in the *Odyssey* of the Phlegraean Fields, inhabited by the Laestrygones, terrifying giants who hurled rocks at Odysseus's ships to prevent them from landing, an obvious reference to volcanic eruption, which is again echoed precisely in the voyage of the Irish Saint Brendhan in the fifth century A.D. during his travels near volcanic Iceland. According to ancient mythology, the monstrous Titans were buried under the Phlegraean Fields after their defeat by the gods of Olympus. Their violent attempts to free themselves caused shaking of the Earth and fiery outbursts or volcanic eruptions.

Not far from Cumae, at the mouth of the Bay of Naples, is the volcanic island of Ischia, settled by Greek colonists as early as the eighth century B.C.[6] Here over a thousand early tombs have been found. As mentioned by Livy, Pindar, Timaeus, and Strabo, Mount Epomeo volcano on the island was frequently active in antiquity, and one of these eruptions, which took place about 350 B.C., destroyed a town on the island and may have contributed to the move of the new Greek colony to the mainland at Cumae. Of this eruption The Greek Strabo wrote in his *Geography*:

> Pithekoussai [Ischia] was once inhabited by Eretrians and Khalkidians and although they lived there in prosperity thanks to the fertility of the soil and to the working of gold, they abandoned the island owing to an insurrection; later they were driven away by earthquakes and eruptions, the sea and boiling waters...hence the myth according to which Tifeo [Typhon] lies in abeyance under the island, and when he moves his body, flames and water spurt out and even tiny islands full of boiling water. Timaeus too, tells that the ancients narrate many marvelous things about Pithekoussai and that only shortly before his time the mountain called Epopeus [Mount Epomeo] at the center of the island, shaken by earthquakes, spat out flames and drove the part between the mountain and the water into the open sea; and the part of the earth reduced to ashes was thrown high into the sky, to fall again onto the island like a turbine and the sea withdrew for three stadia, but not long after withdrawing, it returned and flooded the island with its current; thus the fire on the island was extinguished but the deafening noise was such that the people on the mainland ran from the coast into the hinterland of Campania. The thermal springs on the island are noted for curing sufferers from stones.

Yet Mount Etna always commanded the greatest interest among the ancients. In 693 B.C. an eruption on the south flank of the volcano produced lavas that flowed over the city of Catania. Later, the Greek poet Pindar (522–442 B.C.) called the mountain “the Pillar of Heaven” and said of Etna during an eruption in 475 B.C.: “Where from pure springs of unapproachable fire are vomited from the innermost depths; in the daytime the lava streams pour forth a lurid rush of smoke; but in the darkness a red rolling flame sweeps rocks with uproar to the wide deep sea” (“Pythian Odes,” i, 40). The Greek poet Aeschylus also described an eruption of Etna as well as the “rivers of fire which with ravenous jaws devoured the smiling Sicily” (“Prometheus Vinctus,” 368). The Greek historian Thucydides recorded another eruption of the volcano in 425 B.C., and Diodorus Siculus, writing around 60 B.C., referred to an eruption in 396 B.C., that ravaged the country between Taormina and Catania, arresting the march of the Carthagian general Mago along the coast of Sicily (Diodorus, xiv, 59).

As mentioned earlier, Empedocles from Agrigentum in Sicily took a great interest in Etna, and probably witnessed an eruption in 475 B.C when lava flows reached the coast at Ognina. This philospher became particularly fascinated by volcanoes and spent his last years living near the summit of Etna, observing and contemplating it. Empedocles had a strong following in Sicily, and was regarded as a deity by his disciples. To fulfill this claim and provide his congregation with proof of his divinity, he decided to disappear from the face of the Earth, as if he had ascended to the heavens. Legend has it that Empedocles secretly jumped into the crater of Etna while the volcano was in eruption as a fanatical act of immolation, motivated by his desire to be declared a god who had ascended to Heaven. As shown in Salvator Rosa’s seventeenth-century painting, Empedocles chose the erupting crater as a foolproof secret crematorium, where no trace of his earthly remains could be detected (Figure 4.3). It is said, however, that one of his bronze sandals fell off at the crater’s edge in the fatal leap, as depicted in the painting. It was later found by his followers, who recognized the fraud and realized

4.3. Empedocles leaping into the crater of Etna. The Greek philosopher Empedocles lived in Sicily and studied the activity of Mount Etna. He had novel ideas about volcanic activity, but legend has it that he took his life by secretly jumping into the crater to convince his pupils that he had ascended to heaven. Drawing in brown ink and chalk by Salvator Rosa, ca. 1666; Galleria Palatina, Palazzo Pitti, Florence.

that their spiritual leader had not ascended to Heaven after all. In *Paradise Lost* (III, 469), the English poet John Milton also mentions this fatal leap:

> He who to be deemed
> A god, leaped fondly into Etna flames,
> Empedocles.

Etna's activity was also documented by Romans after their conquest of Sicily in 264 B.C. An eruption on Etna's south flank in 122 B.C. was particularly destructive and the damage so severe that Rome offered the citizens of Sicily relief in the form of remitting the tithe for ten years after the eruption to aid in recovery. That event was one of a series of eruptions in the period 141 to 10 B.C., when volcanic activity in Etna was particularly high, with at least fifteen documented eruptions.

The eruption of Etna in 44 B.C. was an extremely violent event and may have produced a major stratospheric volcanic cloud that spread throughout the northern hemisphere, judging from the evidence of sulfuric acid deposition on the Greenland ice cap from this time. The eruption led to darkening of the sun in Rome at the time of the death of Julius Caesar, and this phenomenon was directly associated with the eruption by contemporary writers such as Virgil (*Aeneid* iii, 570–577):

> Nor was the fact told by the sun alone:
> Earth, air and seas, with prodigies were signed,
> And birds obscure, and howling dogs divined.
> What rocks did Etna's bellowing mouth expire
> From her torn entrails; and what floods of fire!

Ovid also wrote of Etna in *Metamorphoses*, as "glowing with its sulfurous furnaces, will not always be fiery, nor was it always so." He compared the volcano to a living beast, "with many passages and orifices that breathe out flames, then when the beast moves, it changes the channels through which it breathes out the flames, closing some holes, while opening up another."

From the ninth to the eleventh century A.D., Sicily was in Arab and Saracen hands, and their writers refer to Etna as *Gibel Uttamat*, the "mountain of fire," which gave rise to Etna's alternate name of *Mongibello* among modern Sicilians. Mongibello is still the name for the middle cone of Etna, which starts at about 1800 m elevation and ends at the foot of the summit cone.

As the American writer Timothy Ferris (1988) has pointed out, the Romans embraced useful aspects of technology, revered authority, excelled in law, yet theirs was a nonscientific culture.[7] They had considerable curiosity about natural objects, but it resulted mainly in numerous compilations rather than a display of creative intellectual force. Their inclination toward the natural world was practical, reflected also in their ideas about the causes of volcanic activity, based in part on the writings of the Greek philosophers and in part on their own observations of Mount Etna, the Aeolian Islands, Vesuvius, and the craters of the Phlegraean Fields.

Among the earliest Romans to discuss volcanic action was the poet Lucretius (Titus Lucretius Carus 95–52 B.C.), who presented a picture of the entire physical universe in his great poem *De Rerum Natura* (On the Nature of Things). Lucretius was a devoted follower of Epicurus, who in turn had studied under Democritus. Consequently, much of the poem is an exposition of the scientific doctrines of his Greek masters. It is in the sixth book of the poem that he presents his views of several phenomena of external nature, including earthquakes and volcanic eruptions.[8] On the nature of earthquakes, Lucretius had two explanations: collapse or cave-ins, and the force of winds trapped inside the Earth, adopting from Democritus the concept of violent winds within the Earth originally advanced by Anaxagoras. Lucretius wrote:

> Come now and grasp the explanation
> That underlies earthquakes. And first of all,
> Be sure to think that underneath, as well
> As on it, earth is amply fitted out
> With windy caves: it harbors many lakes
> And many caves in its bosom, and cliffs
> And fallen rocks. You must also consider
> That many covered rivers underneath
> The earth's back roll their waves along with force
> And tumbling submerged stones. On every side
> The facts insist that the earth's upper surface
> Is like this. So, with such things underneath
> Linked up and stationed there, the earth above
> Trembles, shaken by cave-ins down below
> When time has undermined vast caverns.
> Indeed, whole mountains fall with huge concussions
> And ripples from them crawl a good distance...
>
> Another cause of earthquakes is a gale from one quarter
> That swoops through subterranean corridors
> And leans with great pressure against the sides
> Of caves deep underground; the earth lunges
> In the direction of the headlong wind
> Pressing against it. Houses built up high
> On the land above threaten to tilt over...

When he turns to volcanic phenomena, Lucretius uses Etna as his prime example. The detailed description of an eruption of the volcano suggests that he may have witnessed its activity and may also have resided in Sicily at one time. The concept of furiously raging wind storms within a cavernous Earth was central to his view of an eruption. He considered Mount Etna hollow within, containing a hot wind that heated the surrounding rocks. The fire from these rocks was swept up, emitting ashes, stones, and black smoke from the volcano:

> I will explain now what the reason is
> For the fires that breathe out from Mount Etna's jaws
> From time to time, with such great turbulence.
> The flaming storm, aroused for a destruction
> That is no mere mishap, first overwhelms
> The fields of the Sicilians, and then
> Draws all the neighboring people to look
> At it. They see the smoky sky, glowing
> With sparks all over, and their trembling hearts
> Fill with anxiety about the state
> Of new conditions Nature is devising.

In seeking an explanation for the volcanic process, he asserted that disease and pestilence is inflicted on the Earth from time to time, causing it to shudder and erupt:

> The seeds of many things are in the body.
> This earth and sky of ours carry enough
> Disease and ill to let a grave sickness
> Gain strength. So we must think that from all space
> The seeds of ills can keep on streaming in
> On our whole earth and sky, and the result
> Is that the earth can suddenly shudder
> When struck by them. Devouring eddies course
> Through lands and sea; Etna's fires stream out,
> The sky bursts into flames...

Lucretius also proposed a mechanical model of Etna's activity, in which air, wind, sand, and stones are pumped into the caverns beneath the volcano by the ceaseless action of the ocean tides and waves:

> Let me tell now
> Of the ways irritated flames suddenly
> Breathe out of Etna's mighty furnaces.

> The whole mountain is hollow at its base
> Supported on its flinty vaults; of course
> The wind and air is everywhere around
> In these grottoes: the air becomes stirred up
> And turns to wind. When that has gotten hot
> It heats up all the stones around it, raving
> Wherever it touches, and heats the ground
> And strikes up hot fire that rapidly
> Rises up high and hurls itself on out
> Through the straight passages of Etna's jaws.
> This fire projects its blaze a long distance
> And tosses sparks a long distance, billowing
> Out thick and murky smoke at the same time.
> It thrusts up stones of such amazing weight
> That you cannot doubt what the whirling force
> Of this breathy wind is. The sea, besides,
> Breaks hard against the mountain's base, and sucks
> The surf back with a pull. The mountain's caves
> Reach out far into the sea—air and wind,
> We must admit, enter Etna this way, and stuff
> Of water, sand and stones roll in from the ocean,
> To be breathed out again through the volcano.
> There the winds are forced to raise the flame
> And roll the stones up from beneath, toss up
> The clouds of sand. For at the top there are
> The "craters" as they call them in the Greek,
> The feature we call Etna's jaws and mouth.

A contemporary of Lucretius was the Greek historian Diodorus Siculus, a native of Agyrium at the foot of Mount Etna. In his writings he mentions volcanoes and volcanic rocks of the Phlegraean Fields, Vesuvius, Etna, and the island of Lipari. "The whole region was named Phlegraean, from the culminating point, which is now called Vesuvius, bearing many indications of having emitted fires in ancient times." He also records that shortly before Hamilcar the Phoenician reached Etna in 394 B.C. the volcano had displayed intense activity, and that the coast at the foot of the mountain was covered in lava, forcing his army to make a detour around it.

The most practical of the Romans was the great architect Vitruvius (ca. 70–30 B.C.), who served as a military engineer under both Julius Caesar and Augustus and wrote the classic work *De Architectura* in his old age. He noted that there must be fires below Vesuvius because of the steam and hot water issuing from the ground, and cited evidence of former eruptions of molten rock that poured over the surrounding countryside. He was also the first to document the unique properties of volcanic ash, or pozzulano, which will harden even underwater when mixed with lime and water, turning into cement. Vitruvius described an experiment that throws much light on the thinking of the Greeks (and some of the early Romans). Taking a bronze *eolipile* (a hollow bronze ball with a small opening through which water is poured), he placed it on a fire. When the water began to boil, out came a strong blast of wind, generated by the heat.

Strabo attributed the elevation and subsidence of land to the action of central fires in the Earth in his *Geographica*. The fires explained the origin of volcanoes, which he regarded as safety valves, providing means of escape for the winds or fiery vapors pent up within the Earth, winds that would otherwise cause earthquakes. He described Etna, Vesuvius, and the Vulcano and Lipari Islands and the great subterranean explosions that accompanied their eruptions.[9] In his time Vesuvius had not been active, but the perceptive Strabo deduced from the character of the rocks on the summit, which were similar to those thrown out from Etna, that Vesuvius was indeed a dormant volcano.

Although Strabo offered other excellent observations on volcanoes and their products, he made no progress in accounting for their origin or behavior. He thought that rock was set on fire by the friction of intensely compressed air in very narrow places in the Earth. Thus, his attribution of volcanic activity to the force of winds pent up within the Earth was borrowed directly from Aristotle and Lucretius. Yet his idea of volcanoes as a safety valve of the Earth was to persist in natural philosophy well into the eighteenth century.

Seneca's Internal Combustion

Lucius Annaeus Seneca (ca. 2 B.C.–A.D. 65) wrote on volcanism in his work on natural philosophy: *Questiones Naturales*, around A.D. 62 (Figure 4.4). Born of a wealthy family in Spain, Seneca became a playwright, philosopher, and finally a highly influential political adviser at the court of Emperor Nero.[10] His works were widely read, but most of his ideas were essentially those of Aristotle's—with an important exception: his view of volcanic activity. He attributed volcanic eruptions in part to the movement of winds within the Earth, struggling to break out to the surface, thus in part following Aristotle. But his most notable and truly original contribution to theories of volcanism is his proposal that the heat liberated from volcanoes is derived from the combustion of sulfur and other flammable substances within the Earth—an idea that was to have many adherents into and even beyond the Middle Ages.[11] His claim was that great stores of sulfur and other combustible substances existed in cavities within the Earth, and that when the great subterranean wind rushed through these regions, frictional heating would set these fuels on fire. In his discussion of hot springs and thermal waters Seneca first makes reference to sulfur and other combustible materials as a potential heat source, proposing that: "Water contracts heat by issuing from or passing through ground charged with sulfur." He then extends this process to explain the blasts from a volcanic eruption:

> We must recognise, therefore, that from these subterranean clouds blasts of wind are raised in the dark, what time they have gathered strength sufficient to remove the obstacles presented by the earth, or can seize upon some open path for their exit, and from this cavernous retreat can escape toward the abodes of men. Now it is obvious that underground there are large quantities of sulfur and other substances no less inflammable. When the air in search of path of escape works its tortuous way through ground of this nature, it necessarily kindles fire by the mere friction. By and by, as the flames spread more widely, any sluggish air there may be is also rarefied and set in motion; a way of escape is sought with a great roaring of violence.

In his view, there were actually a vast number of fires burning at depth within the Earth, some of which occasionally burst forth. "Who can doubt, for instance, that wind gave birth to Thera and Therasis, and to the younger island which even in our own time we have seen spring up in the Aegean Sea?" In this reference Seneca noted that the internal fire is neither extinguished by the weight of the overlying ocean, nor prevented from rising to great heights above sea level. He pointed out that the quantity and nature of the fire in volcanoes were such that even the sea could not quench it:

> Why, according to Posidonius' account, when an island rose in the Aegena Sea long ago in our forefathers' days, the sea was lashed into foam for a long time previously and sent up smoke from its depths. At last fire was emitted, not continuously, but in flames shooting out at intervals, after the fashion of thunderbolts, just as often as the fervent heat of what lay below had overcome the weight of water above it. By and by boulders were thrown up and rocks, part of them still unimpaired, which the air had thrust out before their calcination, part of them corroded by the fire and changed to light pumice; at last the cone of a blasted mountain issued from the waves. Subsequently, there was an addition to its height, and the rock grew in extent into an island.

In addition to the work of Seneca, the most significant writing on volcanism in the Roman world is to be found in poetry. The Latin poem *Aetna*, possibly written between A.D. 63 and 79, is of considerable importance in the history of volcanological thought. The poem's authorship has long been a subject of debate, but its likely author is Lucilius Junior, who had ample opportunity to observe the Mount Etna volcano in his position as Nero's Procurator of Sicily, and he was indeed familiar with volcanic regions, having been born in the vicinity of Vesuvius.[12] He was also a friend and disciple of Seneca, and his authorship of the poem is supported in Seneca's letters. Unlike many of his contemporaries, the poet has a low opinion of mythological explanations for volcanic phenomena and sought to present naturalistic explanations for Earth processes, especially concerning the action of volcanoes. In this regard the poet breaks with the tradition of other Roman scholars, who distinguished in general between *Natura* or nature, and *Terra*, the Earth. Terra was benign, and gave in abundance to humans through agriculture. Natura was the world, the creation and the creator, but Natura is often in conflict with the environment, threatening and even harmful. Thus when Etna erupts, Elder Pliny stated that "nature savagely threatens the lands with fire." Volcanic eruptions were not part of the workings of the Earth, but a rather dark, and inexplicable side of nature,

4.4. The Roman scholar Seneca was perhaps the first to propose that the heat of volcanic action was derived from the combustion of sulfur, bitumen, and other organic substances in the Earth. He also documented the great earthquake near Vesuvius in A.D. 62, the precursor of the A.D. 79 eruption. Eighteenth-century engraving, based on a Roman bust (author's collection).

acting through the medium of fire.[13] The *Aetna* poem is 645 lines and it is found in fourteen surviving manuscripts, the oldest from the tenth century. In it, Lucilius attacks mythological interpretations of volcanic activity:

> Let none be deceived by the fictions poets tell
> That Aetna is the home of a god
> That the fire gushing from her swollen jaws is Vulcan's fire
> And that the echo in that cavernous prison
> Comes from his restless work.

The poet maintained that the Earth is not solid, but has numerous caverns and passages, derived from the Earth's creation when large blocks of rock fell together and left voids or open spaces between them. Some of these channels or passages may have resulted from erosion caused by air, water, and hot vapors flowing through the Earth. He pointed to grottoes, caves, springs, and underground rivers as ample evidence for such deep hollows and passages. Heat, he argues, is more intense and powerful when in action in a confined space within the Earth, with the bellows-like action of winds in subterranean furnaces giving rise to volcanoes. Moving air, or wind, is thus the primary cause of volcanic heat, with fire the result. He proposed several forces as the cause of this wind, such as winds absorbed from the atmosphere by the high peak of Etna, and rockfalls within the Earth, causing violent motion of the internal air. In his view, however, the flames of Etna were fueled by a combustible substance, such as liquid sulfur, oily bitumen, or alum, and the lava stone (*molaris lapis)* is the source of the fire, melting repeatedly to produce the fire until it is exhausted and turns to volcanic ashes:

> A blast of burning sand pours out in whirling clouds.
> Conspiring in their power, the rushing vapors
> Carry up mountain blocks, black sand and dazzling fire.

Ovid also wrote of the transformations of the Earth's surface, ascribing them to pent-up winds. In *Metamorphoses*, he described the phenomenon at Troezene, where the violence of the winds, imprisoned in their dark caverns within the Earth, heaved up the ground like a bladder and made a prominent hill that still endures. Philostratus the Elder (ca. A.D. 190) discusses subterranean passages in the Earth, with fires breaking out through volcanoes: "If one wishes to speculate about such matters, the island of Sicily provides natural bitumen and sulfur, and when these are mixed by the sea, the island is fanned into flame by many winds, drawing from the sea that which sets the fuel aflame." Nonnus of Panopolis in Egypt (A.D. fifth century) may be the last writer of antiquity who wrote on volcanic phenomena in the *Dionysiaca*, in which he discusses the legend of Typhon.

5

The Plinian Eruption of Vesuvius in A.D. 79

Fortune favors the brave.

Pliny the Elder, ca. A.D. 70

When the Greeks colonized Italy, they named the volcanic region west of Naples the Phlegraean Fields (burning fields), which must have been in direct reference to the volcanic activity in the area. And in 16 B.C. the architect and military engineer Vitruvius remarked: "Let it be recorded, that heats in antiquity grew and abounded under Mount Vesuvius, and thence belched forth flame round the country." The larger region that includes both the Phlegraean Fields and Mount Vesuvius is known as Campania, famed among the Greeks and Romans for its fertility and benign climate. The Romans knew that they resided in a volcanic region, but the awakening of long-dormant Vesuvius took them by surprise. "We have heard that Pompeii, the very lively city in Campania, where the shores of Surrentum, and Stabiae and Herculaneum meet in a lovely, gently retreating inlet from the open sea, has been destroyed by an earthquake which also struck the entire vicinity. This occurred in winter, at a time which our forefathers always held to be free from such perils....The region had never before been visited by a calamity of such extent, having always escaped unharmed from such occurrences and having therefore lost all fear of them. Part of the city of Herculaneum caved in; the houses still standing are in ruinous condition." Thus wrote Seneca, in *Natu-*

rales Quaestiones (Natural Questions) of the earthquake of February 5, A.D. 62 in southern Italy. That year saw the first stirring of the foundations of Vesuvius in historical time and also marked the birth of Pliny the Younger, who was to witness, survive, and chronicle the cataclysmic eruption of the volcano seventeen years later. It was an eruption that would claim the life of his uncle, Pliny the Elder, and bury the cities of Pompeii and Herculaneum in a deep layer of volcanic ash, with much of the population and all their art treasures as well.

After a century of peace during the Julio-Claudian dynasty, the Roman empire became embroiled in unrest and civil war from A.D. 67 to 69, following Nero's death.[1] On December 20, A.D. 69 the armies of Flavius Vespasian entered the city of Rome in three columns. After meeting bitter resistance, they killed the emperor Vitellius and threw his body into the Tiber. Thus began the Flavian dynasty, which was to rule Rome for twenty-seven years. Its first emperor was Caesar Vespasianus Augustus II (Figure 5.1 *left*). Vespasian was universally liked and regarded as a good ruler who reorganized the empire, settled the Jewish War, and captured Jerusalem.[2]

Vespasian's popularity was evident in Pompeii, where an altar and temple were erected in his honor in A.D. 74 on ruins of the sanctuary of Genius Agusti, destroyed in the A.D. 62 earthquake. Vespasian fell ill with fever in Campania and died on June 23, A.D. 79, exactly

5.1. At the time of the eruption of Vesuvius in A.D. 79, Titus (right) had taken over as Emperor of the Roman state, following the death of his father, Emperor Vespasian (left) who reigned from A.D. 69 to 79. Vespasian died on June 23, A.D. 79, only two months before the eruption. The bust of Vespasian is from the Ny Carlsberg Glyptotek (Copenhagen) and that of Titus from the Vatican Museum (Rome).

two months before the eruption of Vesuvius. His son Titus, already a co-ruler in the last years of his father's reign, then assumed the title Imperator Titus Caesar Vespasianus Augustus III. Although physically like his father, Titus' temperament and education were very different (Figure 5.1 *right*). Titus grew up in the court of Claudius and had lived under Nero as one of the "gilded youths" close to the emperor. He was a typical product of the Neronian age in culture and tastes and came to the throne with a bad reputation: dissolute, violent in his exercise of official functions, and a target of accusations that he had poisoned his father.[3] Yet on becoming emperor, Titus displayed a surprising change; it was said he laid aside one mask to assume another. The Vespasian-type government continued and he became regarded as a moderate and generous ruler. The eruption of Vesuvius was the most memorable event of his reign. Graffiti discovered in a latrine during the excavation of Herculaneum proves that Apollinaris, his imperial physician, and, by implication, also the Emperor himself, were in there less than a month before the eruption.[4]

It is known that Titus hastened to Campania on receiving news of the eruption and its widespread destruction and returned again the following year. He appointed two "curatores restituendae Campaniae" to organize aid and reconstruction in the region and these important appointments were drawn by lot from those of consular rank. Titus provided the necessary funds from his own resources and from the property of the many people who had died in the eruption without heirs. Contemporary accounts indicate that refugees from the devastated areas around Vesuvius fled to the neighboring towns of Capua, Nola, Neapolis, and Surrentum. The shrewd Titus gave special privileges to the towns that had offered assistance to the refugees.

When Titus returned to Campania in A.D. 80 to oversee the reconstruction he had ordered after the volcanic disaster, more calamity struck that was to divert some of his resources back to Rome. A fire and an epidemic that same year ravaged the imperial city. These misfortunes, so closely following the eruption, were viewed by some as a bad omen for his reign. The Jews, however, viewed the eruption as divine retribution for the emperor's role in the Jewish War and the capture of Jerusalem by the Romans. His reign was a short one, and before Titus could continue his work in Campania, he died on September 13, A.D. 81, a victim of fever like his father.

The eruption of Vesuvius in A.D. 79 is intimately connected to Pliny the Elder, and it is indeed because of the death of this great polymath and elder statesman during the eruption that we have a vivid contemporary account of that disaster. Caius Plinius Secundus was born of noble parentage in A.D. 23 at Lake Comum (Lake Como) in the foothills of the Alps in northern Italy (Figure 5.2). During the fifty-six years of his lifetime many notable events happened. Jerusalem was taken and destroyed by his friend and comrade-in-arms Vespasian and his son Titus; Saints Peter and Paul were martyred in Rome; London came into being as a Roman settlement; an ocean route was opened through the Red Sea to India, and Agricola sailed for the first time around Britain. Pliny the Elder was the most important reporter of Roman science, recording thousands of facts and superstitions in his

5.2. The Roman natural historian Pliny the Elder took part of the Roman fleet on a rescue and research mission across the Bay of Naples during the eruption of Vesuvius in A.D. 79 and perished in the attempt. Engraving by André Thenet, 1684 (author's collection).

monumental work *Historia Naturalis*. Pliny moved in court circles in Rome during the Principate of Caligula and he saw military service in Germany where his patron Pomponius Secundus was governor. During his careeer he also served the Roman Empire in Spain, where he became interested in mining. In this era, the spirits of the wicked dead were consigned to the Inferno, deep below the Earth's surface. The enlightened and hard-headed Pliny, however, noted how strange it was that the miners who dig deep into the Earth had never encountered these infernal regions. He was familiar with many rock formations and also recognized that obsidian is a natural glass. Not much is known about his life, but a revealing letter, which his nephew and heir Pliny the Younger wrote to Baebius Macer, gives us a delightful sketch of his uncle's personality and work habits:

> I was delighted to hear that your close study of my uncle's books has made you wish to possess them all. Since you ask me for a complete list, I will provide a bibliography, and arrange it in chronological order, for this is the sort of information also likely to please scholars.
>
> "Throwing the Javelin from Horseback"—one volume; a work of industry and talent, written when he was a junior officer in the cavalry.

> "The Life of Pomponius Secundus"—two volumes. My uncle was greatly loved by him and felt he owed this homage to his friend's memory.
>
> "The German Wars"—twenty volumes, covering all the wars we have ever had with the Germans. He began this during his military service in Germany, as the result of a dream; in his sleep he saw standing over him the ghost of Drusus Nero, who had triumphed far and wide in Germany and died there. He committed his memory to my uncle's care, begging him to save him from the injustice of oblivion.
>
> "The Scholar"—three volumes divided into six sections on account of their length, in which he trains the orator from his cardel and brings him to perfection.
>
> "Problems in Grammar"—eight volumes; this he wrote during Nero's last years when the slavery of the times made it dangerous to write anything at all independent or inspired.
>
> "A Continuation of the History of Aufidius Bassus"—thirty-one volumes.
>
> "A Natural History"—thirty-seven volumes, a learned and comprehensive work as full of variety as nature itself.
>
> You may wonder how such a busy man was able to complete so many volumes, many of them involving detailed study; and wonder still more when you learn that up to a certain age he practiced at the bar, that he died at the age of fifty-five, and throughout the intervening years his time was much taken up with the important offices he held and his friendship with the emperors. But he combined a penetrating intellect with amazing powers of concentration and the capacity to manage with the minimum of sleep...Before daybreak he would visit the Emperor Vespasian (who also made use of his nights) and then go to attend his official duties. On returning home, he devoted any spare time to his work.... He made extracts of everything he read, and always said that was there was no book so bad that some good could not be got out of it.... In the country, the only time he took from his work was for his bath, and by bath I mean his actual immersion, for while he was being rubbed down and dried he had a book read to him or dictated notes. When traveling he felt free from other responsibilities to give every minute to work; he kept a secretary at his side with book and notebook, and in winter saw that his hands were protected by long sleeves, so that even bitter weather should not rob him of a working hour.... It was this application which enabled him to finish all those volumes, and to leave me 160 notebooks of selected passages, written in a minute hand on both sides of the page, so that their number is really doubled.

The *Historia Naturalis* is Pliny the Elder's largest and best-known work. In compiling this encyclopedic work, he consulted no fewer than 2000 volumes, obtained from one hundred principal authors. It has been said that, unlike the Greeks, no Romans were original scientists. What Saint Augustine said of the Roman scholar Varro can apply equally well to Pliny: that he read so much that it was a marvel he ever had time to write anything, and wrote so much that it was difficult to see how he found time to read. Pliny was possessed with a mania for acquiring information and he searched the great libraries of the time for books on science, most of which have since been lost to us, either burnt in the great fire of Alexandria in Egypt or destroyed when the Roman Empire fell. But in the end, his mania for looking for fresh avenues of knowledge led to his death at Stabiae south of Vesuvius, during the A.D. 79 eruption of the volcano.

Pliny the Elder was familiar with Campania, a beautiful region that supplied wines famous throughout the Roman world, and it is evident from his writings that he had visited its cities on many occasions. He mentions ancient Neapolis, a colony of the Chalcidians, and also Herculaneum and Pompeii, "not far from Vesuvius." The thirty-fifth book of his *Historia Naturalis* contains descriptions of paintings that were actually discovered when the ruins of Herculaneum and Pompeii were excavated. Nowhere in his writings, however, is there an indication that he was aware of the volcanic nature of Vesuvius. In fact, although he listed much information about volcanoes in his *Historia Naturalis*, he expressed no views on volcanic processes or what makes volcanoes erupt. Pliny did discuss an economic aspect of volcanic activity, with respect to the mining of sulfur: "Among other kinds of 'earth' the one with the most remarkable properties is sulfur, which exercises great power over many other substances. Sulfur occurs in the Aeolian islands between Sicily and Italy, which we have said are volcanic, but the most famous is the island of Melos. It is there dug out of mine shafts and dressed with fire."[5]

Later research has shown that Mount Vesuvius had a record of volcanic activity long before A.D. 79, but curiously the volcano had been dormant during the Classic era—and thus had little or no effect on Roman ideas of volcanism. The activity of Vesuvius is cyclic, with very large explosive eruptions occurring at intervals of many hundreds to a thousand years. One of these large events is the so-called Avellino eruption, which occurred in the Bronze Age, about 3340 years ago, coinciding closely with the great Minoan eruption of Thera.[6] The Avellino eruption spread ash fallout widely to the east of Vesuvius, and hot pyroclastic flows and surges must have devastated all life on the volcano's slopes, extending up to 15 km north of Vesuvius. This eruption was followed by a period of dormancy or only minor activity, which lasted 1130 years, until A.D. 79. In the soils around Vesuvius, sandwiched between the prominent layers of pumice and ashfall from these two major eruptions, at least two ash layers provide evidence of intervening explosive eruptions of moderate intensity, one around 1000 B.C. and the other in the seventh century B.C., but no records exist about these events.

Historical records do indicate that an eruption occurred in Vesuvius in 217 B.C.,[7] when the Roman Silius Italicus (A.D. 26–101) wrote: "Vesuvius also thundered, hurling flames worthy of Etna from her cliffs; and the fiery crest , throwing rocks up to the clouds, reached to the trembling stars." During the second Punic War (218–201 B.C.) Hannibal made his famous march across the Alps and crushed the Roman armies in Italy. Again, according to Silius, in 215 B.C. "Hannibal is shown Mount Vesuvius, where fire has eaten away the rocks at the summit, and the wreckage of the mountain lies all around, and the discharge of stones seeks to rival the death dealt by Etna."

Large volcanic eruptions are preceded by a variety of symptoms that can be felt or observed at the surface months or years before eruption begins, and the eruption of Vesuvius in A.D. 79 was no exception. These symptoms may include the swelling or inflation of the volcano and the surrounding land, earthquakes, increased geothermal or hot spring activity, rise or fall of the groundwater table, and increased emissions of volcanic gas. Minor phreatic, or steam, explosions also typically occur shortly before the main eruption. All of these phenomena are linked to the upward flow of magma within the Earth's crust, either from a deep source to a reservoir at 5 to 10 km (3 to 6 mi) below the volcano, or out of the magma reservoir toward the surface via a narrow conduit.

Some early information about Vesuvius dates from the first century B.C., with a glimpse of the volcano linked with the fate of the slave Spartacus. In 73 B.C. a revolt of gladiators and slaves led by Spartacus posed a serious threat to the cities around Vesuvius. Spartacus was from Thrace and apparently a man of superior character and intellect. No contemporary accounts of his life exist, but some of his deeds were recorded in the second century A.D. by Plutarch and the historian Appian. He served as a Roman soldier in his home country, but deserted and became a brigand.[8] Recaptured and sold into slavery in the gladiator school in Capua near Vesuvius, he organized a breakout with seventy other gladiators, seeking shelter in the crater of Vesuvius, which provided a natural fortress. There the band was besieged by Claudius Glaber, who was sent by the Praetor of Rome to mop up this trouble. According to Plutarch,

> There was only one way up this hill and that was narrow and difficult and closely guarded by Glaber; in every other direction there was nothing but sheer cliffs. The top of the hill was covered with wild vines which the gladiators twisted into strong rope-ladders long enough to reach down the cliff-face. They all got down safely by means of these ladders except one man who stayed at the top to deal with the arms, and he, once the rest had got down, dropped the arms down to them and then descended last and reached the plain safely. The Romans knew nothing of this and thus the gladiators were able to get around behind them and throw them into confusion by an unexpected attack, first routing them and then capturing their camp. And now they were joined by numbers of herdsmen and shepherds of these parts, all sturdy men and fast on their feet.

The tale of Spartacus contains meager but very important information about the topography of Vesuvius some 150 years before the great eruption.[9] The chroniclers of the rebellion note that in earlier days the volcano was covered by a dense forest famous for its wild boars. By the first century B.C. the primeval forest had given way to the axe and the plough. Writing at the time of Augustus, in the first century A.D., Strabo identified Vesuvius as an extinct volcano. "Above these places lies Vesuvius, the sides of which are well cultivated, even to the summit. This is level, but quite unproductive. It has a cindery appearance for the rock is porous and of a sooty color, the appearance suggesting that the whole summit may once have been on fire and have contained craters, the fires of which died out when there was no longer anything left to burn."

A painting of Vesuvius from this time that shows the pastoral aspect of the slumbering giant has survived in Pompeii, (Figure 5.3). The volcano is represented as a single, steep peak, its lower slopes covered by a network of vineyards, with trees and bushes growing almost to the summit, where the slopes were too steep for cultivation but supplied firewood and land for grazing. The painting also features Bacchus, clad in a tunic and covered with bunches of grapes. It is one of the earliest works of landscape art showing a volcano.

On February 5, A.D. 62 Vesuvius gave the first signs of returning to activity when a large

5.3. Before the eruption of A.D. 79, Vesuvius was covered with forests, vineyards, and other vegetation up to the summit crater, as shown in this wall painting, found in the House of the Centenary in Pompeii during the excavation. The figure draped in bunches of grapes is the god Dionysius (Bacchus), and the snake represents the household gods. Archaeological Museum of Naples.

earthquake, centered near Pompeii, rocked the area. Seneca wrote an account of this earthquake and other pre-eruption phenomena under the heading *De Terrae Motu* (Earthquakes) in Book VI of *Natural Questions*:

> Lucilius, my good friend, I have just heard that Pompeii, the famous city in Campania, has been laid low by an earthquake which also disturbed all the adjacent districts. The city is in a pleasant bay, back aways from the open sea, and bounded by the shores of Surrentum and Stabiae on one side and the shores of Herculaneum on the other; the shores meet there. In fact, it occurred in days of winter, a season which our ancestors used to claim was free from such disaster. This earthquake was on the Nones of February, in the consulship of Regulus and Verginius. It caused great destruction in Campania, which had never been safe from this danger but had never been damaged and time and again had got off with a fright. Also, part of the town of Herculaneum is in ruins and even the structures which are left standing are shaky. The colony of Nuceria escaped destruction but still has much to complain about. Naples also lost many private dwellings but no public buildings and was only mildly grazed by the great disaster; but some villas collapsed, others here and there shook without damage. To these calamities others were added: they say that a flock of hundreds of sheep was killed, statues were cracked, and some people were deranged and afterwards wandered about unable to help themselves. The thread of my proposed work, and the concurrence of the disaster at this time, required that we discuss the causes of these earthquakes.
>
> Yet certain things are said to have happened peculiar to this Campanian earthquake, and they need to be explained. I have said that a flock of hundreds of sheep was killed in the Pompeian district. There is no reason you should think this happened to those sheep because of fear. For they say that a plague usually occurs after a great earthquake, and this is not surprising. For many death-carrying elements lie hidden in the depths. The very atmosphere there, which is stagnant either from some flaw in the earth or from inactivity and the eternal darkness, is harmful to those breathing it. Or, when it has been tainted by the poison of the internal fires and is sent out from its long stay it stains and pollutes this pure, clear atmosphere and offers new types of disease to those who breathe the unfamiliar air.
>
> I am not surprised that sheep have been infected —sheep which have a delicate constitution—the closer they carried their heads to the ground, since they received the afflatus of the tainted air near the ground itself. If the air had come out in greater quantity it would have harmed people too; but the abundance of pure air extinguished it before it rose high enough to be breathed by people.
>
> In several places in Italy a pestilential vapour is exhaled through certain openings, which is safe neither for people nor for wild animals to breathe. Also, if birds encounter the vapour before it is softened by better weather they fall in mid-flight and their bodies are livid and their throats swollen as though they had been violently strangled.
>
> As long as this air keeps itself inside the earth, flowing out only from a narrow opening, it has power only to kill the creatures which look down into it and voluntarily enter it. When it has been hidden for ages in the dismal darkness underground it grows into poison, becomes more deadly by the very delay, becomes worse the more sluggish it is. When it has found a way out it lets fly the eternal evil and infernal night of gloomy cold and stains darkly the atmosphere of our region.

From Seneca and other sources it is evident that the earthquake was strongest in Pompeii and Herculaneum but affected a larger area, including Nucera and Naples, where the Gymnasium collapsed.[10] In Pompeii, the Vesuvius and Herculaneum Gates to the city fell,

5.4. A marble relief in a shrine in the house of Lucius Caecilius Jucundus in Pompeii shows the effects of the A.D. 62 earthquake. The relief, an accurate account of the earthquake's destruction, shows on the top far left the tottering monumental arch, then the collapse of the Capitolium flanked by equestrian statues; on the right is the altar of Tellus (Earth). The lower relief depicts the Castellum Aquae (Pompeii waterworks) and two mules narrowly escaping the collapsing Vesuvius Gate.

and aqueducts and waterworks were broken, including the lead pipes that served as plumbing in residential houses. Herculaneum suffered even greater earthquake damage, but restoration proceeded much faster in the richer city and was mostly complete by the time of the A.D. 79 eruption. Yet earthquakes are common in Campania and people probably took this one in stride. One prominent banker in Pompeii, Lucius Caecilius Jucundus, commemorated the event with two amusing marble reliefs in the shrine of the household gods in his house that depict the crashing of arches and columns and the heaving of the ground, with two equestrian statues suddenly coming to life as the riders struggle to regain balance (Figure 5.4). The Roman Senate sent aid for restoration in Campania; it was still in progress seventeen years later when the volcano dealt the final blow.

Seneca's reference to the death of hundreds of sheep by mysterious poisoning on the slopes of Vesuvius is perhaps the best evidence that the volcano was already returning to life in A.D. 62 and that the earthquake was related to the volcano. Increased emission of volcanic gases commonly occurs before eruptions, including large quantities of carbon dioxide. Such gases are frequently lethal to people and livestock, and near Hekla volcano in Iceland, for example, flocks of sheep are often killed as they wander into pools of the invisible but suffocating carbon dioxide, which, because of its higher density, gathers in depressions and valleys. The dangers of this invisible killer were well documented in the town of Vestmannaeyjar on Heimaey island south of Iceland in 1973, where large quantities of carbon dioxide gas flowed out of the volcano and the ground surrounding it during an eruption. The gas collected in a ground layer, often so thin that seagulls could safely walk along the deserted ash-covered street, with their heads above the gas layer, whereas the shorter sparrows and snowbuntings keeled over, suffocated.

Two years after the Campanian earthquake, in the spring of A.D. 64, Nero gave a public concert in the theater in Naples. Obsessed with his artistic merits, he still did not dare display them on the stage in Rome and instead chose the more liberal Naples, "the Greek city." During his performance an earthquake struck again, but Nero continued as if nothing had happened. According to Suetonius and Tacitus, the theater collapsed, however, as soon as the performance was over and the audience had left. Descriptions indicate that this earthquake had its origins under the volcanic island of Ischia to the west, and probably not in Vesuvius.

In Campania most of the year A.D. 79 was uneventful until June 24, when news came of the death of Emperor Vespasian and the rise of his son Titus. In Pompeii a statue of Titus was quickly erected in the Temple of Augustus in honor of the new Emperor. In mid-August, earth tremors began anew. The shocks were small but frequent and caused only minor damage. By August 20 the earthquake tremors increased in strength and noises like distant thunder were heard. And at the same time the springs ceased to flow and wells dried up.

In August of A.D. 79, Pliny the Elder was in command of the Roman fleet at Misenum, 32 km (20 mi) across the Bay of Naples from Vesuvius. This excellent natural harbor was well protected but within easy reach of the Tyrrhenian Sea.[11] Nearby Puteoli was the most important city in Italy for the imperial economy, with 150,000 tons of Egyptian wheat imported annually and stored in huge granaries built during the reign of Augustus. The strategic importance of the Bay of Naples has not changed over the ages; it is now the headquarters of the Mediterranean naval fleet of NATO.

Pliny the Elder had been joined that summer at Misenum by his sister and her seventeen-year-old son, Pliny the Younger, and a Spanish friend. During the two days of the eruption, the family was to experience grave danger, which led to Pliny the Elder's death and a narrow escape for Pliny the Younger and his mother. Although the death of this important Roman must have raised questions at the time, there is no evidence that the events were documented until A.D. 103–107, in two letters written by Pliny the Younger.[12] At the request of the historian Cornelius Tacitus, Pliny wrote down his recollections of the circumstances of his uncle's death during the eruption in the first letter:

> Thank you for asking me to send you a description of my uncle's death so that you can leave an accurate account of it for posterity; I know that immortal fame awaits him if his death is recorded by you. It is true that he perished in a catastrophe which destroyed the loveliest regions of the earth, a fate shared by whole cities and their people, and one so memorable that it is likely to make his name live forever: and he himself wrote a number of books of lasting value: but you write for all time and can still do much to perpetuate his memory. The fortunate man, in my opinion, is he to whom the gods have granted the power either to do something which is worth recording or to write what is worth reading, and most fortunate of all is the man who can do both. Such a man was my uncle, as his own books and yours will prove...
>
> My uncle was stationed at Misenum, in active command of the fleet. On 24 August, in the early afternoon, my mother drew his attention to a cloud of unusual size and appearance. He had been out in the sun, had taken a cold bath, and lunched while lying down, and was then working at his books. He called for his shoes and climbed up to a place which would give him the best view of the phenomenon. It was not clear at that distance from which mountain the cloud was rising (it was afterwards known to be Vesuvius); its general appearance can best be expressed as being like an umbrella

pine, for it rose to a great height on a sort of trunk and then split off into branches, I imagine because it was thrust upwards by the first blast and then left unsupported as the pressure subsided, or else it was borne down by its own weight so that it spread out and gradually dispersed. Sometimes it looked white, sometimes blotched and dirty, according to the amount of soil and ashes it carried with it. My uncle's scholarly acumen saw at once that it was important enough for a closer inspection, and he ordered a boat to be made ready, telling me I could come with him if I wished. I replied that I preferred to go on with my studies, and as it happened he had himself given me some writing to do.

As he was leaving the house he was handed a message from Rectina, wife of Tascius whose house was at the foot of the mountain, so that escape was impossible except by boat. She was terrified by the danger threatening her and implored him to rescue her from her fate. He changed his plans, and what he had begun in a spirit of inquiry he completed as a hero. He gave orders for the warships to be launched and went on board himself with the intention of bringing help to many more people besides Rectina, for this lovely stretch of coast was thickly populated. He hurried to the place which everyone else was hastily leaving, steering his course straight for the danger zone. He was entirely fearless, describing each new movement and phase of the portent to be noted down exactly as he observed them. Ashes were already falling, hotter and thicker as the ships drew near, followed by bits of pumice and blackened stones, charred and cracked by the flames: then suddenly they were in shallow water, and the shore was blocked by the debris from the mountain. For a moment my uncle wondered whether to turn back, but when the helmsman advised this he refused, telling him that Fortune favors the brave and they must make for Pomponianus at Stabiae. He was cut off there by the breadth of the bay (for the shore gradually curves round a basin filled by the sea) so that he was not as yet in danger, though it was clear that this would come nearer as it spread....

Meanwhile on Mount Vesuvius broad sheets of fire and leaping flames blazed at several points, their bright glare emphasized by the darkness of night. My uncle tried to allay the fears of his companions by repeatedly declaring that these were nothing but bonfires left by the peasants in their terror, or else empty houses on fire in the districts they had abandoned. Then he went to rest and certainly slept, for as he was a stout man his breathing was rather loud and heavy and could be heard by people coming and going outside his door. By this time the courtyard giving access to his room was full of ashes mixed with pumice-stones, so that its level had risen, and if he had stayed in the room any longer he would never have got out....the buildings were now shaking with violent shocks, and seemed to be swaying to and fro as if they were torn from their foundations. Outside on the other hand, there was the danger of falling pumice-stones, even though these were light and porous; however, after comparing the risks they chose the latter. In my uncle's case one reason outweighed the other, but for the others it was a choice of fears. As a protection against falling objects they put pillows on their heads tied down with cloths.

Elsewhere there was daylight by this time, but they were still in darkness, blacker and denser than any ordinary night, which they relieved by lighting torches and various kinds of lamps. My uncle decided to go down to the shore and investigate on the spot the possibility of any escape by sea, but he found the waves still wild and dangerous. A sheet was spread on the ground for him to lie down, and he repeatedly asked for cold water to drink. Then the flames and smell of sulphur which gave warning of the approaching fire, drove the others to take flight and roused him to stand up. He stood leaning on two slaves and then suddenly collapsed, I imagine because the dense fumes choked his breathing by blocking his windpipe which was constitutionally weak and narrow and often inflamed. When daylight returned on the 26th—two days after the last day he had seen—his body was found intact and uninjured, still fully clothed and looking more like sleep than death.

This letter deals primarily with his uncle's death and only incidentally mentions the natural phenomena during the volcanic disaster and the dangers facing Pliny the Younger and his mother. These other factors clearly aroused Tacitus' interest for he requests further information, to which Pliny replies:

> So the letter which you asked me to write on my uncle's death has made you eager to hear about the terrors and hazards I had to face when left at Misenum, for I broke off at the beginning of this part of my story. Though my mind shrinks from remembering...I will begin.
>
> After my uncle's departure I spent the rest of the day with my books, as this was the reason for staying behind. Then I took a bath, dined, and then dozed fitfully for a while. For several days past there had been earth tremors which were not particularly alarming because they are frequent in Campania: but that night the shocks were so violent that everything felt as if it were not only shaken but overturned. My mother hurried into my room and found me already getting up to wake her if she were still asleep. We sat down in the forecourt of the house, between the buildings and the sea close by. I don't know whether I should call this courage or folly on my part (I was only seventeen at the time) but I called for a volume of Livy and went on reading as if I had nothing else to do. I even went on with the extracts I had been making. Up came a friend of my uncle's who had just come from Spain to join him. When he saw us sitting there and me actually reading, he scolded us both—me for my foolhardiness and my mother for allowing it. Nevertheless, I remained absorbed in my book.
>
> By now it was dawn, but the light was still dim and faint. The buildings around us were already tottering, and the open space we were in was too small for us not to be in real and imminent danger if the house collapsed. This finally decided us to leave the town. We were followed by a panic-stricken mob of people wanting to act on someone else's decision in preference to their own (a point in which fear looks like prudence), who hurried us on our way by pressing hard behind in a dense crowd. Once beyond the buildings we stopped, and there we had some extraordinary experiences which thoroughly alarmed us. The carriages we had ordered to be brought out began to run in different directions though the ground was quite level, and would not remain stationary even when wedged with stones. We also saw the sea sucked away and apparently forced back by the earthquake: at any rate it receded from the shore so that quantities of sea creatures were left stranded on dry sand. On the landward side a fearful black cloud was rent by forked and quivering bursts of flame, and parted to reveal great tongues of fire, like flashes of lightning magnified in size....
>
> Soon afterwards the cloud sank down to earth and covered the sea; it had already blotted out Capri and hidden the promontory of Misenum from sight. Then my mother implored, entreated and commanded me to escape as best I could—a young man might escape, whereas she was old and slow and could die in peace as long as she had not been the cause of my death too. I refused to save myself without her, and grasping her hand forced her to quicken her pace. She gave in reluctantly, blaming herself for delaying me. Ashes were already falling, not as yet very thickly, I looked round: a dense black cloud was coming up behind us, spreading over the earth like a flood. "Let us leave the road while we can still see," I said, "or we shall be knocked down and trampled underfoot in the dark by the crowd behind." We had scarcely sat down to rest when darkness fell, not the dark of a moonless or cloudy night, but as if the lamp had been put out in a closed room. You could hear the shrieks of women, the wailing of infants, and the shouting of men; some were calling their parents, others their children or their wives, trying to recognize them by their voices. People bewailed their own fate or that of their relatives, and there were some who prayed for death in their terror of dying. Many besought the aid of the gods, but still more imagined there were no gods left, and that the universe was plunged into eternal darkness forever more. There were people, too, who added to the real perils by inventing fic-

titious dangers: some reported that part of Misenum had collapsed or another part was on fire, and though their tales were false they found others to believe them. A gleam of light returned, but we took this as a warning of the approaching flames rather than daylight. However, the flames remained some distance off; then darkness came on once more and ashes began to fall again, this time in heavy showers. We rose from time to time and shook them off, otherwise we should have been buried and crushed beneath their weight. I could boast that not a groan or cry of fear escaped me in these perils, had I not derived some poor consolation in my mortal lot from the belief that the whole world was dying with me and I with it.

At last the darkness thinned and dispersed into smoke or cloud; then there was genuine daylight, and the sun actually shone out, but yellowish as it is during an eclipse. We were terrified to see everything changed, buried deep in ashes like snowdrifts. We returned to Misenum where we attended to our physical needs as best we could, and then spent an anxious night alternating between hope and fear. Fear predominated, for the earthquakes went on, and several hysterical individuals made their own and other people's calamities seem ludicrous in comparison with their frightful predictions. But even then, in spite of the dangers we had been through and were still expecting, my mother and I had still no intention of leaving until we had news of my uncle.

These two letters, considered a masterpiece of writing, are most likely based on the young man's conversations with the surviving sailors and friends from Stabiae. There is no reason to doubt his remarkable account of the events on August 24 and 25, which can in fact be reconciled to a high degree with the geologic evidence in the volcanic deposits, although several passages are open to more than one interpretation due to differences between the several surviving Latin texts and obscure phrasing.[13]

The account of Pliny the Elder's death also became a source of contention among classicists.[14] Pliny the Younger describes the death scene at Stabiae on the morning of the second day of the eruption as follows: "The flames and smell of sulfur which gave warning of the approaching fire, drove the others to take flight and him to stand up. He stood leaning on two slaves and then suddenly collapsed, I imagine because the dense fumes choked his breathing." Although the original text is quite specific about the effects of the dense fumes, it has been repeatedly suggested that Pliny the Elder died from a heart attack. However, the evidence in the letters is best interpreted as indicating death by suffocation. Pliny repeatedly asked for cold water to drink because of the sulfurous fumes in Stabiae. Finally, as the cloud of fumes became denser, his often inflamed windpipe became further obstructed, leading to suffocation.

Surely the most imaginative and fantastic writing regarding Pliny's death stems from excavations south of Pompeii. During a private excavation by Gennaro Matrone in 1899–1902, a large villa was discovered at Fondo Bottaro, with an adjoining row of twenty shops, or *tabernae*.[15] The villa stood on the south side of the estuary of the Sarno River and probably overlooked the harbor of Pompeii. Its large peristyle garden was enclosed by thirty fluted columns with doric capitals and contained an unprecedented abundance of statuary, including the well-known bronze Matrone Hercules statuette, resting on a rock of Nocerian tufa.[16] Frescoes of high quality adorn most walls of the villa and have been attributed to the style of painting dating from the time of Emperor Tiberius (A.D. 14–37). The row of shops must have lined the major road or highway between Pompeii and Stabiae. Many sold fishing equipment, while one was a *thermopolium* or snack shop, where hot drinks were served. The seaside location of the villa and its shops was further confirmed by the discovery in 1858 of a Roman *liburnica* (a fast sailing-vessel of the brigantine type) during excavation for the foundation of a mill about 100 m from the villa.[17]

When the volcanic deposit was excavated from the portico in front of the shops, human skeletons were discovered in front of and within almost every taberna. Nine skeletons alone were discovered at the thermopolium, accompanied by a great many valuables, such as jewelry, coins, silverware, and objects in bronze, glass, and terra-cotta. No fewer than 2162 coins of gold, silver, and bronze were found in the villa. The report of the original excavator, Matrone, describes the discovery of the skeleton of an old man in front of taberna number five: "laid on his back, inclined, the head, which was almost entire, resting against a pillar."[18] The skeleton bore a triple gold collar of seventy-five links, weighing 400 g (0.9 lb), as well as triple gold bracelets of weight, value, and fine workmanship like that of the collar. At the side lay a finely wrought ivory-hilted sword. Around the old man were grouped the skeletons of people of wealth, as evidenced by handsome gold jewelry about their necks and arms, some of it set with gems, and by gold and silver money. Among them was the skeleton of a physician with his bronze tubes containing medicaments and surgical instruments. There was an enormous bracelet in the form of a serpent with a lion's head—possibly the emblem of an ensign. In the neighborhood of the skeleton of the old man was the skeleton of a 2.1 m (6 ft, 10 in) giant with a bronze lamp formed like a horse's head. Three meters from the old man was the skeleton of a woman, adorned by two collars of gold, one larger than the other, by bracelets and pearl earrings, and carrying gold money; at her side were two skeletons of children, probably her own. Among other interesting objects (found the following year) were the four bronze bar-ends of a litter, pieces of its resting-support, and a bust of Minerva with squared base, which had probably ornamented the front of the litter. Matrone surmised that the skeleton of the old man was none other than Pliny the Elder, and published this opinion in an article in the *Corriere di Napoli* in November 1901. The scene at Fondo Bottaro certainly has many of the ingredients of the death-scene of Pliny the Elder as documented in his nephew's letters: a wealthy patrician and his entourage on flight from the eruption; a ship beached nearby. But the credibility of Matrone's theory is destroyed by two important facts. First, Pliny the Younger is quite specific that his uncle perished at Stabiae, not at the outskirts of Pompeii. Second, he states that his uncle's body was found on the third day after the beginning of the eruption, "in the dress in which he fell, and looking more like a man asleep than dead." Yet the discoveries at Fondo Bottaro give an important glimpse of the fate of people in flight from the volcano south of Pompeii and indicate that a vast number of the population of the city may have perished as they fled during the second phase of the eruption.

Except for brief remarks by the Latin poets Martial and Statius, the first mention of the eruption after Pliny the Younger's letters is the *Epitome*, or Roman History, of Dio Cassius.

Writing in about A.D. 230, or one-and-a-half centuries after the eruption, Dio gives a garbled account of the event, influenced by superstition and mythology:

> In Campania remarkable and frightful occurrences took place; for a great fire suddenly flared up at the very end of the summer. It happened in this way. Mt. Vesuvius stands over against Neapolis near the sea and it has inexhaustible fountains of fire. Once it was equally high at all points and the fire rose from the centre of it; for here only have the fires broken out, whereas all the outer parts of the mountain remain even now untouched by fire. Consequently, as the outside is never burned, while the central part is constantly growing brittle and being reduced to ashes, the peaks surrounding the centre retain their original height to this day, but the whole section that is on fire, having been consumed has in the course of time settled and therefore become concave; thus the entire mountain resembles a hunting theatre—if we may compare great things to small. Its outlying heights support both trees and vines in abundance, but the crater is given over to the fire and sends up smoke by day and a flame by night; in fact, it gives the impression that quantities of incense of all kinds are being burned in it. This, now, goes on all the time, sometimes to a greater, sometimes to a less extent; but often the mountain throws up ashes, whenever there is an extensive settling in the interior, and discharges whenever it is rent by a violent blast of air. It also rumbles and roars because its vents are not all grouped together but are narrow and concealed.
>
> Such is Vesuvius, and these phenomena usually occur there every year. But all the other occurrences that had taken place there in the course of time, however notable, because unusual, they may have seemed to those who on each occasion observed them, nevertheless would be regarded as trivial in comparison with what now happened, even if all had been combined into one. This was what befell. Numbers of huge men quite surpassing any human stature—such creatures, in fact, as the Giants are pictured to have been—appeared, now on the mountain, now in the surrounding country, and again in the cities, wandering over the earth day and night and also flitting through the air. After this fearful droughts and sudden and violent earthquakes occurred, so that the whole plain round about seethed and the summits leaped into the air. There were frequent rumblings, some of them subterranean, that resembled thunder, and some on the surface, that sounded like bellowings; the sea also joined in the roar and the sky re-echoed it. Then suddenly a portentous crash was heard, as if the mountains were tumbling in ruins; and first huge stones were hurled aloft, rising as high as the very summits, then came a great quantity of fire and endless smoke, so that the whole atmosphere was obscured and the sun was entirely hidden, as if eclipsed. Thus day was turned into night and light into darkness. Some thought that the Giants were rising again in revolt (for at this time also many of their forms could be discerned in the smoke and, moreover, a sound as of trumpets was heard), while others believed that the whole universe was being resolved into chaos or fire. Therefore they fled, some from the houses into the streets, others from outside into the houses, now from the sea to the land and now from the land to the sea; for in their excitement they regarded any place where they were not as safe than where they were. While this was going on, an inconceivable quantity of ashes was blown out, which covered both sea and land and filled all the air. It wrought much injury of various kinds, as chance befell, to men and farms and cattle, and in particular it destroyed all fish and birds. Furthermore, it buried two entire cities, Herculaneum and Pompeii, the latter place while its populace was seated in the theatre. Indeed, the amount of dust, taken all together, was so great that some of it reached Africa and Syria and Egypt, and it also reached Rome, filling the air overhead and darkening the sun. There, too, no little fear was occasioned, that lasted for several days, since the people did not know and could not imagine what had happened, but like those close at hand, believed that the whole world was being turned upside down, that the sun was disappearing into the earth and that

the earth was being lifted to the sky. These ashes now did the Romans no great harm at the time, though later they brought a terrible pestilence upon them.

However, a second conflagration, above ground, in the following year spread over very large sections of Rome while Titus was absent in Campania attending to the catastrophe that had befallen that region. It consumed the temple of Serapis, the temple of Isis, the Saepta, the temple of Neptune, the Baths of Agrippa, the Pantheon, the Diribitorium, the theatre of Balbus, the stage building of Pompeii's theatre, the Octavian buildings together with their books, and the temple of Jupiter Capitolinus with its surrounding temples. Hence the disaster seemed to be not of human but of divine origin; for anyone can estimate, from the list of buildings that I have given, how many others must have been destroyed.

From Dio's account we learn that Vesuvius was still active in his time, with minor explosions usually occurring "every year." We also learn that the volcanic ash from the A.D. 79 eruption reached not only Rome, "filling the air overhead and darkening the sun," but also Africa, Syria, and Egypt, and that in Rome the ashfall was blamed for the terrible pestilence that visited the city at this time. Strabo claimed that the ash fallout also caused immense damage in southern Italy.[19] In the region surrounding the great Greek city of Paestum the ashfall diverted the nearby river, which invaded the countryside and the city of Paestum, rendering the cultivated fields into useless swamps and marshlands.

The record of the great eruption of A.D. 79 can be read from the stratigraphy of the volcanic deposits. It is the goal of geology, and its branch of volcanology, to reconstruct natural processes of the past from the character of the deposits they leave on the ground. In this case of the volcanic deposits around Vesuvius have proven a rich source of information and, in fact, have given us a story much more complete and factual than the meager eyewitness accounts. My involvement with this endeavor began in 1979 when, on a casual visit to Pompeii with fellow volcanologist Steven Sparks, we realized that a great and untold story resided in the diverse volcanic layers that buried this city and, in fact, were spread over the entire region of Campania. During the next five or six years, with research funding from the National Geographic Society and the National Science Foundation, I initiated a program with Italian, English, and American colleagues, especially the American volcanologist Steven Carey, to reconstruct the sequence of events during the volcanic disaster as documented in the deposits, not only in Pompeii and Herculaneum, but at a number of excavations all around Vesuvius.[20]

The salient points of our study reveal that the great noontime explosion on August 24 was preceded by a small explosion, probably in the morning or during the night. This minor event may not have alarmed the population of the cities around Vesuvius, but may have been the cause of alarm for the lady Rectina, who sent a plea for help to Pliny the Elder. During the twelve hours following the major noontime explosion, the volcano sent a column of ash and pumice into the stratosphere, to a height of 15 to 25 km (9 to 16 mi) above Vesuvius. We can establish accurately the height of this eruption column from the size of the pumice and rock fragments found in the fall deposit at various distances from the crater.[21] Volcanologists refer to all explosive eruptions of this type, characterized by a high eruption column, as Plinian eruptions, in honor of the Plinys. Material from this gigantic column, which Pliny described as resembling a Mediterranean pine in shape, was carried with the strong stratospheric winds to the southeast, resulting in pumice and ash fallout over Pompeii and other sites on this side of Vesuvius. Within about four hours the accumulation of pumice and ash had made the streets in Pompeii impassable to all but pedestrians, and roofs had collapsed under the weight of the deposit. Although it was midday, the dense volcanic cloud overhead spread throughout the entire sky, blotting out sunlight and leaving the region in total darkness. The fall of pumice and ash was accompanied by stones. These are rock fragments, some the size of an egg or a fist, torn from the throat of the volcano and carried by the steady but jet-like flow of the erupting mixture out of the crater and to a great height of 15 to 25 km (9 to 16 mi). The fall of such stones could prove lethal, but on the whole the Plinian stage of the eruption was not a severe hazard to the Pompeiians. However, being pelted by stones must have created great concern and perhaps many fled, wading in total darkness through the loose pumice layer on the ground, most likely to the south, away from the volcano.

By midnight, when Pompeii was already covered by a layer of pumice about 2 m thick, a sudden change occurred in the nature of volcanic activity that was to prove lethal to the inhabitants first of Herculaneum, then Pompeii. Instead of ejecting the fragmentary volcanic material to the stratosphere, as during the Plinian stage, Vesuvius now generated a fountain of glowing hot ash and pumice that towered several kilometers above the crater. From this fountain ash flows, or glowing avalanches, cascaded down the flanks of the volcano at great speed. Few who have witnessed this most deadly of all volcanic phenomena have lived to describe it. It was first well depicted following the tragic eruption of Mount Pelée on the Caribbean island of Martinique in May 1902 and appropriately given the name *nuée ardente*, or the glowing avalanche cloud. Other names are pyroclastic surges or ash flows. Today they are best known from the deadly eruptions of Mount Saint Helens in the Cascade Range in 1980 and El Chichón volcano in Mexico in 1982. Six such ash flows were generated during the second stage of the Vesuvius eruption, from midnight to daybreak on August 25, of progressively greater extent and destructive potential. The first swept into Herculaneum about midnight, with virtually no warning. The people of Herculaneum had already witnessed the twelve-hour spectacle of the Plinian stage of the eruption, but although located only 6 km (3.7 mi) northwest of the crater, there was no pumice and ash fallout in this city because of the wind direction in the stratosphere, which carried all the ash plume to the south, over Pompeii. The Herculaneans probably had a false sense of security that was violently shattered when the first ash flow raced through the city around midnight and continued out to sea. Although total darkness prevailed, the people would have seen the glowing avalanche speed down the flanks of Vesuvius and toward their city at velocities of more than 100 km per hour. They fled from the volcano toward the waterfront and, caught between the deadly glowing avalanche and the sea, were quickly engulfed in the hot ash, burned, and almost immediately asphyxiated. Excavations of the Herculaneum beach in 1982 and later have revealed the

5.5. Pliny the Younger and his mother at Misenum, A.D. 79. In this 1785 painting by the Swiss artist Angelica Kauffman, Pliny is being warned to flee the impending eruption of Vesuvius, but decides to continue his studies despite the danger. The young man looks up from a volume of Livy's writing as his uncle's Spanish friend urges him and his mother to flee. The artist has set the scene in daylight, although Pliny's letter clearly indicates that this event occurred at night. Oil on canvas, The Art Museum, Princeton University.

remains of hundreds of people in the ash flow deposit, arrested in their death struggle.[22]

Pompeii meanwhile was far enough from Vesuvius to be out of reach of the first deadly surge cloud that had devastated Herculaneum. During the following hours several surges were produced from the crater, each larger and greater in extent. The third one rushed up to the city walls north of Pompeii but did not enter the city. About 8 A.M. the fourth even larger surge, came racing down the flanks of the volcano and overran Pompeii. Anyone moving about in the city—by this time they would have been walking on a thick deposit of pumice fall—was instantly knocked over by the blast and suffocated in the searing hot ash cloud. Today we find the remains of the Pompeiians at the level of the fourth surge layer in Pompeii. The fifth and sixth surge clouds, generated shortly afterwards, were even larger, but passed chiefly over lands already laid to waste. The sixth surge traveled well south of Pompeii and reached the outskirts of the town of Stabiae, where Pliny the Elder had sought refuge at the home of his friend Pomponianus. Here the weak Pliny collapsed in the arms of his slaves and died from suffocation at the distal margin of the sixth surge shortly after daybreak. At the same time Pliny the Younger saw the sixth surge speed across the Bay of Naples and head in

5.6. Early excavation work at Herculaneum. The eighteenth-century excavations were done more in the spirit of a treasure hunt than an archaeological study, but the thick and indurated volcanic deposit restricted most of the work to tunneling at the base of the 20-m-thick layer. The engraving is from J. C. R. Abbé de Saint-Non, *Voyage Pittoresque ou Description des Royaumes de Naples et de Sicile*, 1781–86 (author's collection).

the direction of Misenum, as a great and threatening cloud. This frightening sight was sufficient to cause him and his mother to flee from the region, and head for safety north along the coast toward Cumae (Figure 5.5).

The cities of Vesuvius were now wiped from the face of the Earth and buried to a depth beyond the reach of excavation by the Romans. The deposit over Herculaneum was up to 23 m thick, and Pompeii lay under a blanket of 4 m of pumice. Gradually the memory of the cities faded into oblivion, until their rediscovery in the eighteenth century (Figure 5.6). However, the exterminator Vesuvius remained active. Although historical accounts of its eruptions after the sack of Rome are vague, we know of a large explosive eruption in A.D. 472 that caused ash fallout as far east as Constantinople (Istanbul). The volcano has had a rich and varied history of activity since then. The largest of these was the explosive eruption of 1631, which is discussed further in Chapter 7, but the most recent eruption was in 1944.

6

The Chimneys of Hell

As for the Earth...under it is turned up as it were by fire.

Book of Job

Scholarly work on the nature of the Earth ceased with the end of the Classical period and the fall of the Roman Empire in the Dark Ages (ca. A.D. 400 to 1000), and it can be said that with the rise of Christianity, the European continent suffered a scholarly amnesia during the Middle Ages to about A.D. 1300.[1] Some learning and knowledge of nature inherited from the writers of antiquity survived in monasteries, but little progress was made in philosophical or scientific inquiry—and studies into the nature and origin of the Earth were virtually nonexistent. The impediments imposed by the new religion were many—no opinions were allowed that ran contrary to orthodox beliefs. Scriptural dogma led to a retreat of knowledge on all fronts concerning Earth science, even to a retreat to the concept of a flat Earth.[2] Thus, science fared no better and possibly worse than under the pagans in Rome. It made no difference to the Christian scholar, who was engrossed in spirituality, ascetism, and thoughts of the afterlife, whether the world was round or flat. "To discuss the nature and position of the Earth does not help us in our hope of the life to come" said Saint Ambrose in the fourth century, and for Tertullian the Convert "Curiosity is no longer necessary." These scholars were inherently uninterested in the study of material things.

Instead, religious philosophers of the Middle Ages told their contemporaries to raise their eyes to the heavens and forget about the Earth. In fact, Saint Augustine (A.D. 354–430) banned the entire spectrum of the natural sciences that had been developed in antiquity when he wrote in the fifth century:

> When the question is asked what we are to believe in regard to religion, it is not necessary to probe into the nature of things, as was done by those whom the Greeks call *physici*; nor need we be in alarm lest the Christian should be ignorant of the force and number of the elements—the motion and order and eclipses of the heavenly bodies; the form of the heavens; the species and the nature of animals, plants, stones, fountains, rivers, mountains; about chronology and distances; the signs of coming storms; and a thousand other things which those philosophers either have found out or think they have found out. It is enough for the Christian to believe that the only cause of all created things, whether heavenly or earthly, is the goodness of the Creator, the one true God.

This irrational attitude toward science continued well beyond the Middle Ages. Even by the eighteenth century, most British writers on the philosophy of nature considered that God created the Earth as a habitat for humans; that it had undergone certain stages of evolution since, and would ultimately be transformed or destroyed by Him. This was still manifest in the writings of John Wesley (1703–91), the founder of Methodism, who taught that before sin entered the world, there were no earthquakes or volcanoes. These convulsions of the Earth were simply the "effect of that curse which was brought upon the earth by the original transgression."

Every facet of nature had its place in scriptural dogma, and the role that volcanoes might play was obvious. They were the "chimneys of Hell," and the subterranean fires that issued from them were invariably identified with the place of everlasting torment. Thus, the fiery Mediterranean island of Vulcano was considered by the Italians as the place of punishment for the hated Arian Emperor Theodosius I, and Mount Etna was assigned by the British to Anne Boleyn, the English queen who brought Henry VIII to disavow the pope.

Brave monks in quest of Paradise were the heroes of the Middle Ages, and journeys to Paradise were the most popular topic of medieval literature. The most famous of these travelers was the Irish Saint Brendan the Navigator (A.D. 484–578), a noted churchman and abbot who founded several monasteries in Ireland and made numerous voyages to England, Scotland, and Brittany. During an epic journey in the North Atlantic he encountered a volcanic eruption. Believing that Paradise was to be found in the North Atlantic, Brendan set off westward from Galway with fourteen other monks in search of this Promised Land. At that time Ireland and the British Isles represented the northwestern bounds of the known world—the daring Viking voyages to Iceland, Greenland, and Vinland or North America were still some 400 years in the future. The vessel carrying the Irish monks was a small skin-covered *curragh*: a simple wooden frame covered with tanned oxhide. The voyage, described in the *Navigatio Sancti Brendani Abbatis*, clearly took them to the north, where they encountered icebergs. But perhaps the most awe-inspiring part of their journey came the day they reached a

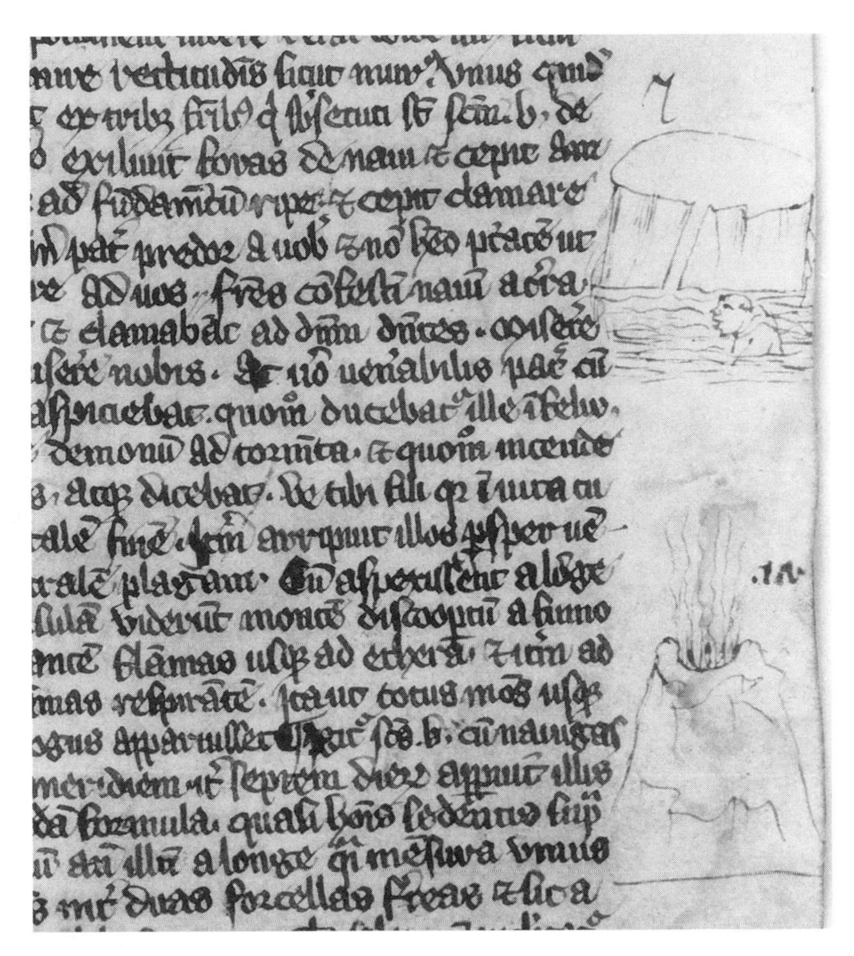

6.1. Saint Brendan's view of a volcanic island in the North Atlantic. The Irish monk may have explored Iceland and the far northern Atlantic more than 300 years before the Vikings. The legend of Saint Brendan's voyage, *Navigatio Sancti Brandani,* was well-known throughout Europe in the early eleventh century. Bodleian Library, Oxford.

rocky, fire-scarred island with neither trees nor grass, where subhuman creatures rushed down to the shore and hurled at their boat red-hot masses of lava and slag, which fell hissing into the sea. They escaped out of range, but the next day saw a volcano on the island with a fiery and smoking summit, evidently the source of the bombardment. There can be little doubt that this volcano was either on Iceland or a volcanic island off its coast where eruptions are relatively common. In one of the mauscripts of the *Navigatio* a delightful sketch of the eruption they witnessed is presented (Figure 6.1). It is possibly the second oldest known pictorial representation of a volcanic eruption, after the Stone Age drawing from Çatal Hüyük in Anatolia, discussed in Chapter 2.

Three hundred years after Brendan's voyage, Iceland was settled by Vikings from Norway and from Norse settlements in the British Isles. The conversion of the Icelanders to Christianity gives an interesting glimpse into their understanding of volcanic processes and reveals an attitude of mixed superstition and clear logic. In the summer of the year A.D. 1000, a fierce debate on the question of conversion from belief in the Nordic gods to Christianity took place at the open-air *Althing*, or parliament, at Thingvellir. The site, within a spectacular volcanic fissure in the active zone of rifting and volcanism in south Iceland, is bounded by vertical basalt lava cliffs that provided excellent acoustics for the parliamentarians, and is floored by a 5000-year-old lava flow with the typical ropy surface texture of basalt lavas. As the debate heated, the adversaries came close to blows with "each man declaring the other outlaw, the Christian men and the heathen men against one another." Arguing came to temporary halt when a messenger came running to the Althing, bringing news of a volcanic eruption in nearby Olfus, and that the new lava flow he reported was threatening to overwhelm the farm of the chieftain Thoroddur. Many of the heathen parliamentarians regarded this as a clear sign of the wrath of the Norse gods against the new religion and began to argue even more vehemently against converting to Christianity: "It is no wonder that the gods are angry at such talk." To this the shrewd politician and realist Snorri Godi, the Speaker of the Althing, replied, as he gazed over the vast prehistoric lava field that covered the valley: "What angered the gods when the lava burnt which we are standing on now?" His simple logic convinced the assembly that volcanic activity was normal and that eruptions were part of Iceland's past and would be so in the future. The resolution to adopt Christianity was passed (Figure 6.2). It is interesting to note the phrase "lava burnt" (*jardeldur* or Earth-fire) is the common term for

6.2. The site of the Icelandic *Althing*, or parliament, in the tenth century was in the volcanic rift valley of Almannagja. When Christianity was proclaimed here in the summer of A.D. 1000, a volcanic eruption occurred nearby. The people saw it as a sign of the anger of the Viking gods, whereas the sage speaker of the Althing, Snorri Godi, pointed out that volcanic eruptions were a normal part of nature's workings (author's collection).

6.3. The crater of Hekla volcano in Iceland was regarded by many Christians in the Middle Ages as the gateway to Hell. Mid-nineteenth-century etching (author's collection).

volcanic activity in Iceland. This in itself implies a process of combustion and may reflect the influence of Seneca's view on the early Icelanders—that volcanic activity was the result of combustion of coal, sulfur, and other volatile matter within the Earth.

By the Middle Ages, the Icelanders had also adopted the Christian view of volcanoes as the abode of the devil and the gateway to Hell, with the volcano Mount Hekla the dominant entrance (Figure 6.3). The fearsome noises that issued from some of their volcanoes were certainly thought to be the screams of tormented souls in the fires of Hell below.

In about 1180 Chaplain Herbert of the Cistercian monastery of Clairvaux in France connected volcanoes to the netherworld, comparing Mount Etna with the awesome Hekla: "The renowned fiery cauldron of [Etna in] Sicily, which men call Hell's Chimney...is affirmed to be like a small furnace compared to this enormous inferno.... Who is there so refractory and unbelieving that he will not credit the existence of an eternal fire where souls suffer, when with his own eyes he sees the fire of which I have spoken?" In the fourteenth century Caspar Peucer also wrote of the fearsome Icelandic volcano: "Out of the bottomless abyss of Heklafell, or rather out of Hell itself, rise melancholy cries and loud wailings, so that these may be heard for many miles around. Coal-black ravens and vultures flutter about. There is to be found the Gate of Hell and whenever great battles are fought or there is bloody carnage somewhere on the globe, then there may be heard in the mountain fearful howlings, weeping and gnashing of teeth."

In medieval Europe the Christian view of the netherworld did not differ greatly from the Hades of the Greeks.[3] Saint Hildegaard (1099–1179) described exactly what happened to the sinners there:

> I saw a well deep and broad, full of boiling pitch and sulfur, and around it were wasps and scorpions, who scared but did not injure the souls of those therein; which were the souls of those who had slain in order not to be slain. Near a pond of clear water I saw a great fire. In this some souls were burned and others were girdled with snakes, and others drew in and again exhaled the fire like a breath, while malignant spirits cast lighted stones at them. And all of them beheld their punishments reflected in the water, and thereat were the more afflicted. These were the souls of those who had extinguished the substance of the human form within them, or had slain their infants. And I saw a great swamp, over which hung a black cloud of smoke, which was issuing from it. And in the swamp there swarmed a mass of little worms. Here were the souls of those who in the world had delighted in foolish merriment.

One of the most detailed—and architecturally complex—views of Hell is given in a guided tour of the Inferno in Dante's *Divine Comedy* (1320). Dante also speculated on the powers that had raised the lands above the primeval ocean, opting for an extraterrestrial force. Action was effected "by way of compulsion by generating vapors, as in some mountain parts," a clear reference to upheaval of the Earth's crust by volcanic activity. He described the Inferno as a deep hole, extending from the Earth's surface in the northern hemisphere and terminating in a pit at the center of the Earth. The hole was formed by the impact of the devil Lucifer and his angels as they struck the Earth in their fall from Heaven. The material displaced by the impact emerged on the opposite side of the Earth, in the southern hemisphere, where it formed an island and the mountain of Purgatory, with the Earthly Paradise at its summit. Fire, of course, features very prominently in the Inferno, but as traditional punishment is used only inside the walls of the Citadel of Dis (or Satan), where in the Sixth Circle the Heretics burn in their fiery graves (Figure 6.4). Based on Dante's description, the young Galileo Galilei later wrote a technical report of the structure of the Inferno as a student thesis (1587).[4]

In 1558 Giovanni Paolo Gallucio also proposed a structure for the interior of the Earth that was in keeping with scriptural dogma. Beneath the outer region of rocks and water, the region where metals were found and earthquakes and other surface or near-surface events occurred, was Limbo, followed by Purgatory; then, in succession, the regions of the fighters, the vainglorious, the gluttonous, the oath takers, the angry, the covetous, the proud, the traitors, and, finally, at the center, Lucifer's realm. This arrangement ensured that the inhabitants of the central region were as far removed as possible from God.

In the Middle Ages, a completely new attitude toward art and nature prevailed. Nature had become but a manifestation of divine power, and contemporary paintings clearly reflect this. Landscape, portrayed principally as a backdrop to a religious or historical scene, was often presented in somewhat abstract form. Mountains appeared as strange twisted peaks, spiny projections rising abruptly from the plain, representing a pictorial tradition dating to

6.4. The fourteenth-century poet Dante Alighieri described the universe of Christian truth, including the Inferno or Hell, in his poem *The Divine Comedy.* Here the authoritative guide Virgil visits the Inferno, inspired by the bubbling mud pits in the volcanic region of the Phlegraean Fields near Naples. Engraving by French nineteenth-century illustrator, Gustave Doré (author's collection).

Hellenistic painting and Byzantine art. In the thirteenth century the Italian scholar Cennino Cennini wrote: "If you wish to paint mountains properly, take some big stones, rough and not cleaned up, and paint them as they are, putting in the light and shade as reason recommends." The mid-eleventh century illuminated commentary on the Apocalypse by Stephanus Garsia of "Rain, hail, fire and blood upon the Earth," is an interesting example of the abstraction of landscape of the time—and shows a phenomenon undoubtedly inspired by a volcanic eruption.[5]

Hell and Purgatory were portrayed in an equally abstract manner—as fiery compartments deep in the Earth, as shown, for example, in the Gothic view of Hell by French painter Enguerrand Charonton in 1454. In the late 1400s, an abstract representation of Hell often emerges in the paintings of Hieronymus Bosch and in the works of Joachim Patinir, such as the painting of *Charon's Boat* in which Charon is shown paddling between Heaven and Hell, with the fires of the underworld in sharp contrast to the shining white lakes of the Elysian Fields.

Another well-known view of Hell appeared in John Milton's poem *Paradise Lost.* Milton traveled in Italy, where he visited Galileo, and on his Italian journey he may even have seen Mount Vesuvius. He describes the fall of Satan and the angels into Hell:

> Hurl'd headlong flaming from th' Ethereal Sky
> With hideous ruin and combustion down
> To bottomless perdition, there to dwell
> in Adamantine chains and penal Fire.
>
> Hell at last
> Yawning receiv'd them whole, and on the clos'd,
> Hell their fit habitation fraught with fire
> Unquenchable, the house of woe and pain.

Milton's Hell is a horrible dungeon containing a burning lake and a fiery deluge that is "fed with ever-burning sulphur unconsum'd." Even so, Satan declares: "Better to reign in Hell, than to serve in Heav'n!" He then sets his demons to work to build the great palace of Pandemonium on the side of a volcano.

The idea of a firestorm within the planet was held by some alchemists as late as the seventeenth century. The pioneer scholar German Johannes Joachim Becher (1635–82) believed, on the basis of certain passages in the Holy Writ, in a great subterranean fire in the center, surely the locus of Hell, a conclusion he supported by the consideration that the damned could not possibly be subject to greater pain and torment than what they would experience in this dreadful place. Becher published his ideas in the *Physica Subterranea* in 1669.

Yet some leading scholars were beginning to challenge the conventional view of Hell, struggling to integrate what they were discovering about the universe with the teachings of the church. The philosopher and atheist Thomas Hobbes (1588–1679) was one of the first to ridicule this theological cosmography of the underworld. René Descartes (1596–1650) also maintained that the material universe was fully separate from the spiritual one, and while he considered the universe as a mechanism operating in accordance with laws set down by God, he denied the existence of supernatural events. The Dutch lens-grinder, philosopher, and theologian Baruch Spinoza (1632–77) went even further and substituted for God "the being of the universe." And Isaac Newton (1642–1727), as shown by his private papers, had gradually abandoned faith in miracles and began to doubt the eternity of Hell. However, many remained firm believers, even applying the scientific method to the study of Hell. In 1714 Tobias Swinden, a member of the Royal Society of London, published *An Enquiry into the Nature and Place of Hell*, in which he showed by calculation, first, that during the span of humanity, the accumulation of souls would have long since overrun any subterranean space, and, second, that there was not sufficient oxygen available to keep the underground fires alive. But he was not about to abandon the idea of Hell. He merely moved it to a new location; instead of a volcanic or subterranean heat source, he presented a scientific and logical proof that the Sun, possessing an eternal and enormous fire and ample space for all the lost souls for eternity, was the site of Hell. The English mathematician William Whiston (1667–1752), Newton's successor in the Lucasian chair at the University of Cambridge, published a study, the *Astronomical Principles of Religion, Natural and Reveal'd* (1717), in which he proposed that Halley's comet was the obvious place for Hell. Later, however, he revised his ideas on Hell, and in *The Eternity of Hell Torments Considered* (1740) adopted the more conventional view, replacing Hell deep inside the Earth. With the Enlightenment by the late 1700s the concept of Hell was considered by most scholars as either irrelevant or preposterous and was attacked directly by such philosophers and scientists of the age as the French writers Denis Diderot (1713–84) and François Voltaire (1694–1778).

The Alchemist's Furnace

The early Middle Ages of western Europe coincided with a remarkable growth of learning in Asiatic countries, in particular in Persian and Arabic cultures, that were in their prime from A.D. 800 to 1100. Contrary to the Christian philosophy of the time, the Koran encouraged the Islamic scholar to practice the mastery of *taffakur*, or the study of Nature. Unlike most learned Europeans, who saw Nature as a vivid illustration of the moral purposes of God and a rational explanation of the phenomena described in the Bible, the Arabs sought knowledge that would give them power over Nature. After the fall of Rome to the Goths in 410 and Alexandria to the Arabs in 643, the knowledge nurtured by the Romans was lost, but Islamic scholars studied, preserved, and elaborated on the classic Greek science now all but forgotten in the West. The most significant contributions at this time were made by the Arabs who practiced their art in Iraq and later in Spain, kept alive the interest in researching the secrets of Nature, and developed the practice of alchemy into chemistry. One of the elements that figured prominently in their studies was sulfur, which was in part mined from volcanoes. Their discovery that sulfur could also give off heat was to generate the idea that volcanic action was also fueled by the combustion of sulfur in the Earth.

Alchemy is only one episode in man's experimentation with various chemicals throughout the ages. It is usually regarded as the attempt to transmute base metals, such as lead, to gold with the aid of the famed "philosopher's stone," or the pursuit of an *elixir vitae* or panacea that would cure the ills of mortal life. This is a great oversimplification because the aims of alchemists were generally much broader, and in pursuing these aims they laid the foundations for chemistry.[6] Perhaps the earliest alchemists were those who experimented with medicinal drugs and metals. Copper was already being smelted in the early Bronze Age (2200 to 700 B.C.), both in Britain and continental Europe. Chinese alchemy dates back to at least 133 B.C., when Taoist alchemists attempted to change cinnabar into gold and prepared a large range of elixirs intended to bring the human body to harmony and perfection with the universe, and even to gain immortality. The later contributions of the Chinese alchemists included the invention of gunpowder and fireworks in the tenth century.

The beginnings of alchemy in the West can be traced to the first century A.D., where the practice flourished in Alexandria, Egypt, for about three centuries. There the Greek philosophers amalgamated alchemist practice with their theories of matter and change. These early alchemists were dedicated to laboratory experimentation, and their work gave rise to empiricism in science and the realization of the importance of experiment to scientific progress. They maintained that the less noble metals grew deep in the Earth, in the womb of mother nature, and that with time they gradually changed into the more noble metals. In fact, mines were sometimes sealed off to allow exhausted seams of ore to "recover" and cause more metals to grow. Some alchemists thought that metals grew because of the mixture of fiery and smoky sulfur compounds with the wetter mercury in the Earth. The early Greek alchemy was, as noted, preserved by the Arabs and recycled to the Latin West in the eleventh century. Although the Arabic alchemists' search for gold and the elixir of life was, of course, doomed to failure, they acquired considerable skill from their experiments and made substantial work in the field of chemistry, whereas their work in the study of the Earth was less notable. The Arabic scholars made their greatest contribution in preserving the knowledge of the Classical Greek and Roman Ages, such as the writings of Aristotle, which had been suppressed and lost in Christian Europe. This led to the rediscovery of Aristotle by European scholars when his works were translated from Arabic into Latin between 1200 and 1225.

The most famous of the Arabic scholars was Abu-Musa-Jabir-ibn-Haiyan, who flourished about 776. His writings, which later appeared in Latin under the name Geber, held that the six known metals differed because they contained different proportions of sulfur and mercury, but sulfur (identified with fire) and mercury (identified with liquidity) were considered the primary elements. Another important Arabic scholar was Ibn Sina, also known as Avicenna (980–1037), a famous physician and translator of Aristotle who wrote the encyclopedic *Book of Healing*, a comprehensive work on philosophy and science laced with Aristotelian doctrine. His principal work, however, was *De Mineralibus*, in which he grouped minerals as stones, sulfur minerals, metals, and salts—or similar to the grouping of Agricola several centuries later. He again put forward the Greek theory that some mountains were formed by the eruption of winds imprisoned under the Earth (Figure 6.5).

After Nonnos of Panopolis in late antiquity, there was a long silence on volcanic phenomena until the Middle Ages, when the ideas of both Aristotle and Seneca on the causes of volcanic activity were revived. Following the translation and introduction of the works of Aristotle and the other Greek scholars of antiquity into Europe, Albertus Magnus of Cologne

6.5. The views of the alchemists concerning processes within the Earth in the early seventeenth century were strikingly similar to contemporary ideas of volcanic action (see Figure 7.3). This copper engraving from *Symbola Aurae Mensae Duodecim Nationeum* (1617) is a representation of the sulfur-mercury theory of metals, where the blending of sulfur (triangular symbol on the left) and mercury (circular symbol on the right) generates the "Hermetic Steam," which in turn leads to the synthesis of the philosopher's stone.

(1205–80) became Aristotle's chief interpreter and perhaps the greatest scientific mind of the Middle Ages. He followed Aristotle closely in describing the action of subterranean winds, which, he proposed, elevated the Earth's crust when they were prevented from escaping to the surface. Volcanoes, he held, were formed by the action of these subterranean winds, setting fire to local accumulations of inflammable material within the Earth, thus combining Aristotle's ideas on wind with Seneca's on combustible matter. Avicenna had also regarded volcanic eruptions as a product of winds imprisoned within the Earth, putting forth this theory in *De Mineralibus et Rebus Metallicis* (ca. 1260). This view was also expressed by the German scholar Konrad von Megenberg (1309–74), who wrote in *Buch der Natur* that earthquakes and volcanoes were the result of winds within caverns in the Earth that forced themselves to the surface. Not all medieval scholars were followers of Aristotle, however. The activity of volcanoes and the abundance of hot springs in Italy led Ristoro D'Arezzo to conclude in 1282 that the interior of the Earth was very hot, and probably an incandescent mass, capable of deforming the overlying crust—essentially the same view that Empedocle put forth in the third century B.C.[7] A very different notion was advocated by the Italian Giovanni de'Dondi in the fourteenth century; he explained that heat in the Earth was a result of subterranean fires produced by the action of celestial rays. Many scholars considered the penetrating power of solar rays very great and also explained several other phenomena by this process, including the development of gold deposits in the Earth. This belief was supported by the claim that gold was found in the greatest abundance in tropical countries where the power of sunlight was greatest.

One of the most important of the later alchemists was the Austrian Philippus Aureolus Theophrastus Bombast von Hohenheim (1493–1541), who rechristened himself Paracelsus. He was educated in medicine, but was also familiar with the mines and metalworkings in Tyrol. Concerned more with the chemical preparation of drugs than with the transmutation of metals, Paracelsus scorned those "who have a golden mountain in their heads before they put their hands into the charcoal."[8] In his time, alchemy was regarded as the study of the universe, whereas chemistry was the art of distillation. He thought that alchemy and medicine were controlled by *archeus* and *vulcanus*, which bring about all chemical processes. It was the general opinion of the alchemists that a great body of fire existed at the center of the Earth, giving off dense clouds of vapors that led to the formation of ore deposits in the Earth's crust. They took as evidence of this fiery region the ejection of steam clouds and ashes from Vesuvius, Etna, and other active volcanoes, but opinions differed about how the fire was lighted or fueled. Writing in 1652, the great German alchemist John Rudolf Glauber considered "an empty space in the Earth's center, where nothing is at rest, into which the powers of all the stars are poured forth and where their mutual reactions give rise to an intense heat." Glauber further stated that if this was not the actual lake of fire prepared for sinners, it must at least be situated somewhere in the immediate vicinity. The Italian scholar Girolamo Fracastoro (1483–1553), in *De Ortu et Causis Subterraneorum*, also regarded the Earth's interior as heated by the combustion of bitumen, coal beds, and sulfur, which produced hot vapors that were forced violently through narrow passages in the Earth, causing earthquakes and volcanic eruptions.

The idea of volcanism as a form of combustion prevailed throughout the seventeenth century. In 1661 the English diarist John Evelyn (1620–1706) wrote in *Fumifugium* of the burning of coal, the clouds of which produce a "horrid smoake" that lies "perpetually imminent" and makes the city of London resemble Mount Etna, Vulcan, Stromboli, and the other volcanic suburbs of Hell. This view was echoed by John Milton in *Paradise Lost*:

> Of thundering Etna, whose combustible
> And fuel'd entrails thence conceiving fire,
> Sublim'd with mineral fury, aid the winds
> And leave a singed bottom, all involv'd
> With stench and smoke.

Etienne de Clave was a French scholar who in 1635 adopted the view of a fire within the Earth that burns as eternally within the planet as the Sun does above the Earth. The inner heat acts through exhalations and vapors, which convey heat from the center to all parts of the Earth. The vapors are derived from rain, part of which seeps to great depths, carrying with it some bitumen, sulfur, and other soluble substances. On encountering the subterranean fires in the hot interior, the liquid is volatilized and returns again to the surface, emerging as a volcanic eruption.

Alchemistic ideas on the origin of the Earth and "the material world," based in part on the Genesis account, were discussed by the doctor of medicine and alchemist Edward Dickinson in 1702.[9] In his scheme, the divine spirit set particles of matter in motion, giving rise to the elements. The total mass of matter then took on a rotary motion, with the heavier particles pushed toward the center, forming a globe. Fiery sparks or *bractaea* penetrated through cracks in the globe and kindled the central fires, which aided in the generation of metals. The concept of internal combustion as a source of volcanic activity was, however, very long-lived among European scholars. Jean Baptiste de Lamarck wrote in *Hyderogeology* in 1802: "Nature ceaselessly produces huge concentrations of combustible matter, which water gradually transports to the bottom of the external crust, where they become the supply for all the Earth's volcanoes."

Although alchemy was practiced into the eighteenth century, its pursuit fell into disgrace, and it is appropriate to consider that with the death of James Price (1752–83), alchemy died as well. Price, a member of the Royal Society of London, firmly believed in the transmutation of metals. It is likely that he was not entirely confident about the validity of his own experimental results because when challenged by Sir Joseph Banks and other fellows of the Society to repeat his experiments, he committed suicide.

7

Renaissance and Earth Science

Diseased nature oftentimes breaks forth
In strange eruptions; oft the teeming earth
Is with a kind of colic pinch'd and vex'd
By the imprisoning of unruly wind
Within her womb; which, for enlargement striving,
Shakes the old bedlam earth, and topples down
Steeples and moss-grown towers.

William Shakespeare, *Henry IV*

The beginning of the Renaissance marked the transition from centuries of intellectual domination of theology during the Middle Ages to a new era of human reason in the Western world. It thus delineates the shift away from divine science to natural science, and with it the first real progress of scientific thought since the time of classical Greece. After centuries of medievalism, a humanist revival began in Europe in the fifteenth century that helped widen mental horizons and paved the way for the coming of modern science.[1] European cities, growing and prosperous, were powered by an upsurge in trade, industry, and technological experimentation, which in turn led to the reawakening of science. Although the upsurge of science in the Renaissance had a solid footing in economic life, it was foremost a revolt of the human mind against the stifling traditions of the Middle Ages. Religious dogma stated that the Earth is stable and unmoving and that we are at the center of the universe. The true beginnings of modern science arose when this dogma was denied by the heliocentric theory of the Polish astronomer Nicolas Copernicus (1473–1543)—the revolutionary idea that the Earth itself moved around the Sun. People now looked at the world in a new light and the Earth sank to a lowlier place among the other planets.

7.1. Nineteenth-century engraving of the volcanic peak of the island of Tenerife, a well-known landmark to navigators of the Atlantic Ocean (author's collection).

Scholars such as Leonardo da Vinci (1452–1519) emerged, who considered observations of nature and experiment as the only true methods of science. Leonardo maintained that the secret to the fountain of knowledge was *saper vedere*, "to know how to see," and was the first to recognize that fossils found in rocks high on inland mountains were the remains of organisms formerly living in ancient oceans.[2] He also realized that large changes had occurred in the Earth's crust during geologic time as a result of uplift and erosion. One cannot help wondering about insights he would have had on the workings of volcanoes had he lived in the volcanic region of Naples instead of northern Italy.

The growth of knowledge about the Earth in the Middle Ages was greatly hampered by the primitive fears of mountains and the wilderness, which were not fully overcome until the Renaissance. In medieval Christian philosophy, our earthly life is no more than a brief and squalid interlude, therefore our natural surroundings need not absorb our attention. There was a mistrust of nature; Saint Anselm in the twelfth century wrote that things were harmful for the soul in proportion to the number of senses they delighted. The Italian poet Francisco Petrarch (1304–74) may have been one of the first to break away from this view and express the desire to escape from the turmoil of cities to the peace of the countryside, the mountains, and nature. Unlike the fear and terror of mountains and forests expressed by most medieval poets, Petrarch writes to a friend: "Would that you could know, with what joy I wander free and alone among the mountains, forests and streams." It is often said that Petrarch was the first person to climb a mountain for its own sake, and to enjoy the view from the top.[3] But after he had enjoyed the splendid view of the Alps from Mount Ventoux for a while, the good

Cistercian monk opened at random a copy of Saint Augustine's *Confessions* and read the following passage: "And men go about to wonder at the heights of the mountains, and the mighty waves of the sea, and the wide sweep of rivers, and the circuit of the ocean, and the revolution of the stars, but themselves they consider not." Petrarch was abashed, angry with himself for enjoying the earthly things, and thus reminded that nothing is wonderful but the soul, he hurried away from the mountain. True pioneers in the age of discovery of Nature included the Swiss philosopher Konrad Gesner (1516–65), who was a tireless traveler and explorer of high mountains, which had long been seen with awe and terror. Solitary, isolated mountain peaks, such as volcanoes, were a source of much interest to such travelers. Perhaps the most typical was the lone volcanic peak on the island of Tenerife in the Atlantic, which was discovered by the Portuguese navigator Vasco de Gama in 1497 (Figure 7.1).

The economic upsurge in Europe in the Renaissance created demand for raw materials mined from the Earth. This practical study of minerals and geologic formations led to new ideas about the Earth and its internal heat, mostly due to the great German mining engineer Georgius Agricola (1494–1555), one of the outstanding figures in the history of geological science. Born Georg Bauer in Germany but trained in medicine in Italy, Agricola practiced medicine in mining towns in Bavaria, where he came into close and intimate contact with the Earth and its minerals. He was later a professor of chemistry at Chemnitz in Saxony. In his great work *De Re Metallica* (On Metals), published in 1546, Agricola maintained that the subterranean fires of the Earth were fanned by the fire spirit (*Spiritus Ignitus*) that became active either when cold presses out the fire, as clouds create lightning, or when vapors compressed in a narrow space are heated by the friction until they finally catch fire, as Aristotle had proposed. Agricola considered that pumice was a stone found in places where subterranean fires were burning or had burned in the past. He based his theory of volcanic action on ideas similar to those of the ancients, emphasizing the role of winds within the Earth as an agent of volcanism and mountain building. The winds gave rise to earthquakes, which sometimes burst the Earth's crust, allowing the wind in turn to escape through volcanoes. He also proposed, like Seneca before him, that the burning of bitumen and sulfur generated subterranean fires or volcanic heat, and that volcanoes acted as the vents for the fires. Thus, Agricola did not believe that the entire central region of the Earth was in a highly heated condition, but that the subterranean fires were confined to regions beneath volcanoes, where the combustible substances were at hand. This view was later adopted by the Neptunists, the group of late eighteenth-century geologists who believed that most rocks were produced by sedimentation or precipitation from sea water and considered volcanism an entirely superficial and minor component of Earth's action.[4]

In seeking a mechanism for the driving forces of volcanic action and earthquakes, the French ceramist and glass-maker Bernard Palissy (1520–90) also proposed in his pioneering work *Discours Admirables* (Admirable Discourses) that volcanoes originated by the action of fires fueled by coal, bitumen, soil, and sulfur, as did the alchemists.[5] Once lighted, the fire boils the water in the surrounding rocks, causing its ascent as hot water or steam, which may lead to upheaval, earthquake, and volcanic action. "I have had no other book than the sky and the earth, which is known to all, and it is given to all to know and to read in this beautiful book. Now, having read it, I have studied earthly things, because I had not studied astrology in order to contemplate the stars," wrote Palissy in 1575.

The Dominican father Valerius Faventies wrote the first treatise on the origin of mountains *De Montium Origine*, published in Venice in 1561. In it he discussed the important role of subterranean fire in mountain builiding, citing Mount Etna and Vesuvius, "a marvel which I would that our countryman Pliny had never coveted the sight of. He got too near in his curiosity and met his death." He described the formation of a volcanic island in the Aegean, citing the Greek philosopher Posidonius (ca. 135–50 B.C.), where "the sea foamed for a while and smoke swept up to the sky, with fire issuing from its rifts. Then rocks rolled up to view and crags appeared, some of them eaten hollow by fire and crushed, and last of all the peak of a burning mountain came into sight, which grew and grew until it reached the size of an island. I will say nothing of the similar work of nature which came to light near Naples." Here he is probably referring first to an eruption in the submarine caldera of Santorini (Thera) in the Aegean, and second to the eruption near Pozzuoli in the Bay of Naples, which built up the volcanic cone of Monte Nuovo in 1538 (Figure 7.2).

The appearance of Monte Nuovo created much fear and astonishment because it

7.2. The 1538 volcanic eruption of Monte Nuovo in the Phlegraean Fields near Naples shown in a contemporary woodcut. Prior to the eruption, the shoreline of the bay of Pozzuoli was uplifted 6 to 7 m and the sea receded 200 to 300 m from the coastline, due to updoming of the Earth's crust above the ascending magma.

emerged virtually overnight from a region of fertile fields along the coast. Simone Porzio, a contemporary Italian writer, described the event:

> The region of Puteoli was shaken by rather intensive earthquakes for two years, but intermittently, so that hardly any house was unshaken and there was general concern over moving their site to some other place. But on the 27 and 28 of September in the year 1538 the earth shook continually throughout the day and night. The sea receded 200 paces, and there the local inhabitants took a great quantitiy of fish and the water was fresh. Then on the 29th day a larger tract of land which lies between the foot of Mt. Gaurus, which is called Monte Barbaro, and the sea near Avernus was seen to push itself up and suddenly take the form of a mountain gradually increasing in size. On this same day, the second hour of the night, this mound of the earth, through a vent, and with a great noise, spewed forth strange fires, pumice, stones, and so great a quantity of dark ashes that it covered the buildings of Puteoli which survive to this day. It covered all vegetation, broke trees, and for six miles around, reduced the hanging grape harvest to ashes. The ash near the center was dry; but farther off, it fell muddy and damp. But it surpasses all wonder that what is viewed around the crater (now called Monte Nuovo) accumulated from pumice and ash to an altitude of more than 1000 paces, and all in one night. There were many vents, of which today two survive, the one next to Avernus, the other in the middle of the mountain. A great part of Lake Avernus was covered by ash. This burning remains to this very day, but with some interruption.

The English biologist the Reverend John Ray (1627–1705), writing in 1693, also assumed that subterranean volcanic fires led to the uplift of mountains, citing the formation of Monte Nuovo. He considered volcanoes connected with earthquakes and, through the heating of "steams and damps," the combustible materials in the hollows of mountains would be set on fire and metals and minerals melted.

Giordano Bruno (1548–1600), a Dominican monk who broke with his order to devote his life to science, pointed out the common location of volcanoes near the sea, and suggested that volcanic activity might be the result of a reaction of sea water with the hot interior of the Earth. The Church would not let go of its grip of Bruno, however, and after eight years in the hands of the Inquisition, he was burned at the stake for heresy in 1600.

It has been said that the last of the Renaissance men actually lived in the seventeenth century. He was Athanasius Kircher (1602–80), a prolific German Jesuit scholar, with a fantastically broad range of interests including all the sciences and most of the arts. This adventurous polymath fled his homeland during the Thirty Years War.[6] He traveled widely on behalf of the Catholic church, visiting in his journeys the volcanic regions of Italy. In 1637 Kircher was chosen as the companion and confessor of Frederick of Hesse during his travels in southern Italy. Here Kircher had the opportunity to approach the jaws of Hell, observing eruptions of Etna and the continuously active island of Stromboli. On reaching Naples, he was determined to climb Vesuvius and found a local guide to lead him to the summit. They climbed at night, and their way was lit by the subterranean fires: "When finally I reached the crater, it was terrible to behold. The whole area was lit up by the fires, and the glowing sulfur and bitumen produced an intolerable vapour. It was just like hell, only lacking the demons to

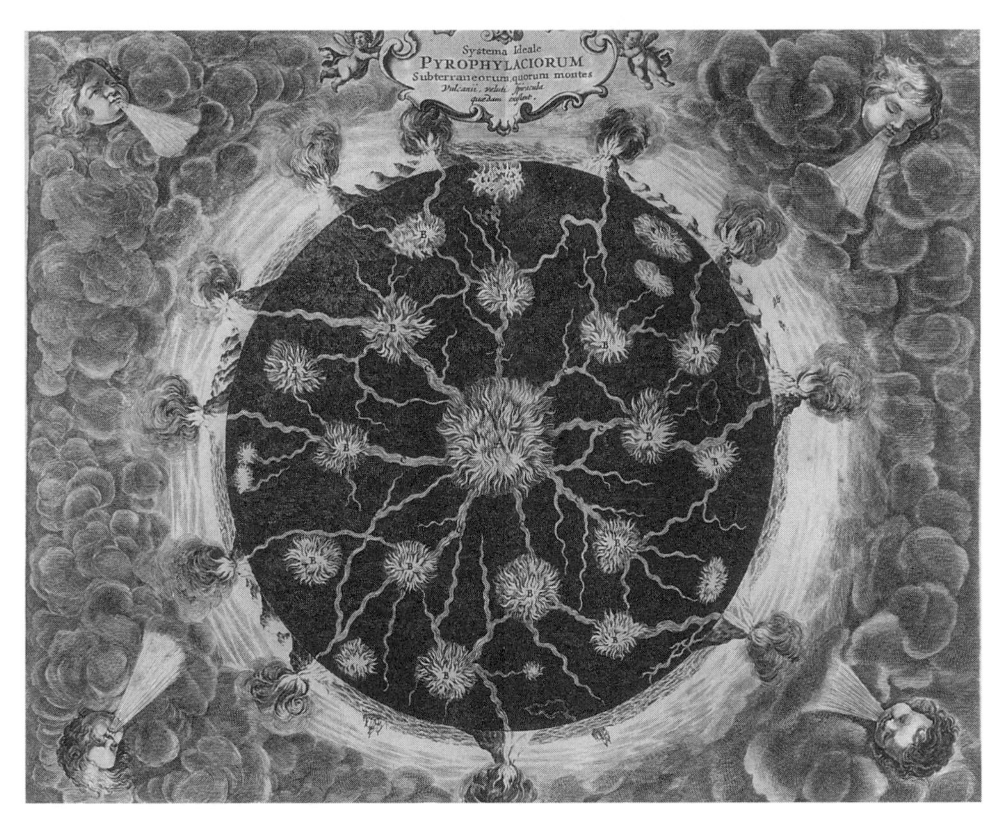

7.3. Athanasius Kircher constructed the first cross section of the Earth in 1665, showing central fires connected by a network of fissures that feed the volcanoes at the surface. From *Mundus Subterraneus* (Amsterdam) in The Royal Society Library, London.

complete the picture." At dawn, Kircher began to explore the crater and summit of the volcano, clambering everywhere he could, descending the crater as far down as possible to observe more closely the volcanic process and the aftereffects of eruption.

In his travels Kircher probably saw more volcanoes than many modern volcanologists, and his visits were to profoundly influence his writings. In 1664 he published a treatise on the Earth in three volumes entitled, "*Mundus Subterraneus, A work in which are exposed the divine workmanship of the underground world, the great gifts of nature there distributed, the form, wonder riches and great variety of all things in the Protean region: the cause of all hidden effects are demonstrated and the application of those which are useful to mankind, by means of various experimental apparatus, and new and previously unknown methods, are explained.*" The book immediately created great interest among Kircher's contemporary scholars in Europe, including Henry Oldenburg, Robert Boyle, Christian Huygens, Benedict Spinoza, and John Locke—all of whom were eager to acquire copies of the thousand-page work.[7]

Kircher took note of miner's reports that temperature increased with depth in the Earth's crust and was probably the first scholar to do so. In the third volume of the work, *Pyrologus*, Kircher developed the idea of a great and inexhaustible fire at the Earth's center, connected by a network of canals *(Pyragogi)* to many smaller bodies of fire *(Pyrophylacia)* at shallower levels; in turn these were connected to volcanoes at the surface. Volcanoes, fumaroles, and hot springs had their origin, ultimately, through the action of the central fire, and the volcanoes were also in part the conduits through which air was drawn into the Earth to vent the flames in the central fire. It is likely that Kircher was influenced here by the idea put forward earlier by the French scholar René Descartes in *Opera Philosophica* (1644) that at the center of the Earth there remains a nucleus of the glowing matter that once composed the whole world. Kircher related this deep internal fire to the four Aristotelian causes; the fire itself as the formal cause; salt, sulfur, niter, bitumen, and other inflammables from the sea as the material cause; wind as the efficient cause; and volcanic vents acting as nozzles of the bellows. It is no coincidence that this potent mixture contained the essential ingredients of gunpowder. Kircher was also impressed by the similarity of the sounds of volcanic explosion and those produced in the firing of a cannon, which, no doubt, led him to attribute volcanic eruptions to the explosion of a mixture of sulfur and niter in the bowels of the Earth.

To illustrate his concept of subterranean fires, Kircher drew a cross section through the globe, with a central fire, and channels or fissures trending upward to the surface, feeding volcanic eruptions—all in an intricate, rococo style (Figure 7.3). Many other beautiful plates illustrate the text and have establilshed Kircher as a pioneer in the use of graphics in the Earth sciences. His cutaway view of Vesuvius (Figure 7.4) shows it as a hollow volcano, with a fire within, in a design exactly like the ovens or furnaces of the alchemists. In a similar view

7.4. Athanasius Kircher drew his scientific view of Vesuvius in 1638, showing a cutout of the volcano's flank to portray his ideas about a raging fire within the mountain. From *Mundus Subterraneus* (Amsterdam) in author's collection.

7.5. An engraving of Mount Etna in eruption in 1637 by Athanasius Kircher. At the volcano's base is the city of Catania, which was overwhelmed by lava flows during the 1669 eruption. From *Mundus Subterraneus* (Amsterdam) in author's collection.

of Etna (Figure 7.5), huge rocks are hurled out of the crater during a volcanic explosion. Kircher was also a good field geologist; he thus recognized the Alban Lake near Rome as an ancient volcanic crater.

In 1669 an English version of *Mundus Subterraneus* was published in London, entitled *The Vulcanoes, or Burning and Fire-Vomiting Mountains, Famous in the World*, and no book did more to stimulate interest in the Earth's interior in the seventeenth and early eighteenth centuries. A mixture of religion, legends about the Earth, and new observations, it stands squarely between the old world of superstition and the new one of scientific observation.[8]

A contemporary of Kircher's was Nicolaus Steno (1638–86), born Niels Stensen in Denmark, of a wealthy family of goldsmiths. Steno was educated in Copenhagen, but carried out his geological studies in Italy in the region around Florence where he was in the service at the court of the Medici as the physician of the Grand Duke Ferdinand II of Tuscany. It was there that he published his great work *Prodromus* (The Forerunner) in 1668. Steno regarded fire in the Earth as an important part of a number of geologic processes, such as uplift, subsidence,

7.6. Seventeenth-century views of Mount Etna showing the volcano's surroundings as well as a detail of the summit region, with its numerous erupting vents (author's collection).

and mountain building.[9] To explain large-scale subsidence of strata, he proposed the collapse of "vast cavities eaten out by the force of fire." He also held that volcanic eruptions were caused by violent escape of burning gases from within the Earth, and were related to the combustion of sulfur and bitumen as suggested by Seneca and others, and believed that volcanic mountains were formed by accumulation of ejected ashes and stones. He was the first to point out that old deposits of volcanic ashes and pumice indicated that they had been formed by ancient volcanic eruptions.

The theory of internal combustion advocated by most medieval scholars as the source of subterranean fires was finally dealt a deathblow by Edward Jorden (1569–1632), an English physician and chemist who sought a new explanation of the generation of metals and the source of the internal heat in the Earth.[10] In the *Discourse of Naturall Bathes and Minerall Waters* (1632), Jorden rejected the notion that the Earth is a hollow and fiery furnace with a universal fire fueled by combustible coal, bitumen, or sulfur. Because all heavy things descend toward a center, he reasoned, the center of the Earth was likely to be denser and more compact than the surface. He further pointed out the fundamental problem regarding the internal fire: it would require a tremendous amount of air to keep it burning. Any flame that is confined without access of abundant air would soon be put out because "fuliginous vapours...choake it if there were not vent for them into the ayre." Such abundance of air could not reach the deeper portions of the Earth to fan the central fire. Instead, Jorden proposed that volcanic regions are underlain at only a shallow depth by fermenting material. As a source of the heat associated with the origin of metals and heating of natural spring waters from within the Earth, he thus sought an explanation in the process of chemical reactions as a basis for a new system of the Earth. Fermentation could take place in the presence of water, which was clearly abundant, deep in the Earth's crust—as miners could attest to—whereas combustion could not proceed in the presence of water. Although his reasoning logically should have ended the hypothesis of combustion as the heat source in the Earth, the idea persisted well into the late eighteenth century, with many influential advocates, among them the "Father of Geology," James Hutton.

The most destructive and largest eruption of the Sicilian volcano Etna occurred in the spring of 1669.[11] This event is of particular importance because of the excellent observations, some even from the summit of the volcano (Figure 7.6), made by the Sicilian natural philosopher and scholar Francesco d'Arezzo (d. 1672). The activity started near the town of Nicolosi at 800 m (2620 ft) above sea level, and soon the twin-peaked cinder cone Monte Rossi was formed. The initial earthquake and lava flows that followed destroyed or damaged over forty towns, displacing between 20,000 to 60,000 people from their homes. After three more days of severe earthquakes, the south flank of the volcano was split asunder by a 12-km (7.4 mi) long fissure that issued copious amounts of lava to the south. Eventually the lava reached the 20-m (66 ft) high city walls of Catania, flowed over them and destroyed a large section of the city (Figure 7.7).

After the eruption d'Arezzo, at the request of his friend Giovanni Alfonso Borelli, agreed to study the lava.[12] When he started up the volcano, fully expecting to find rivers of liquid bitumen and sulfur for he had earlier reviewed the works of Greek and Latin writers on lava flows, he noted: "All say unanimously that the rivers of fire (called seas and oceans by many) that this miraculous mountain has belched forth at various times, to the ruin of Sicilian fields, are nothing else but liquid sulfur and burning bitumen." But his findings were quite contrary to his expectations and led to his demonstrating in a charming manner that lavas could not be liquid sulfur. First he had this to say on the question of odor:

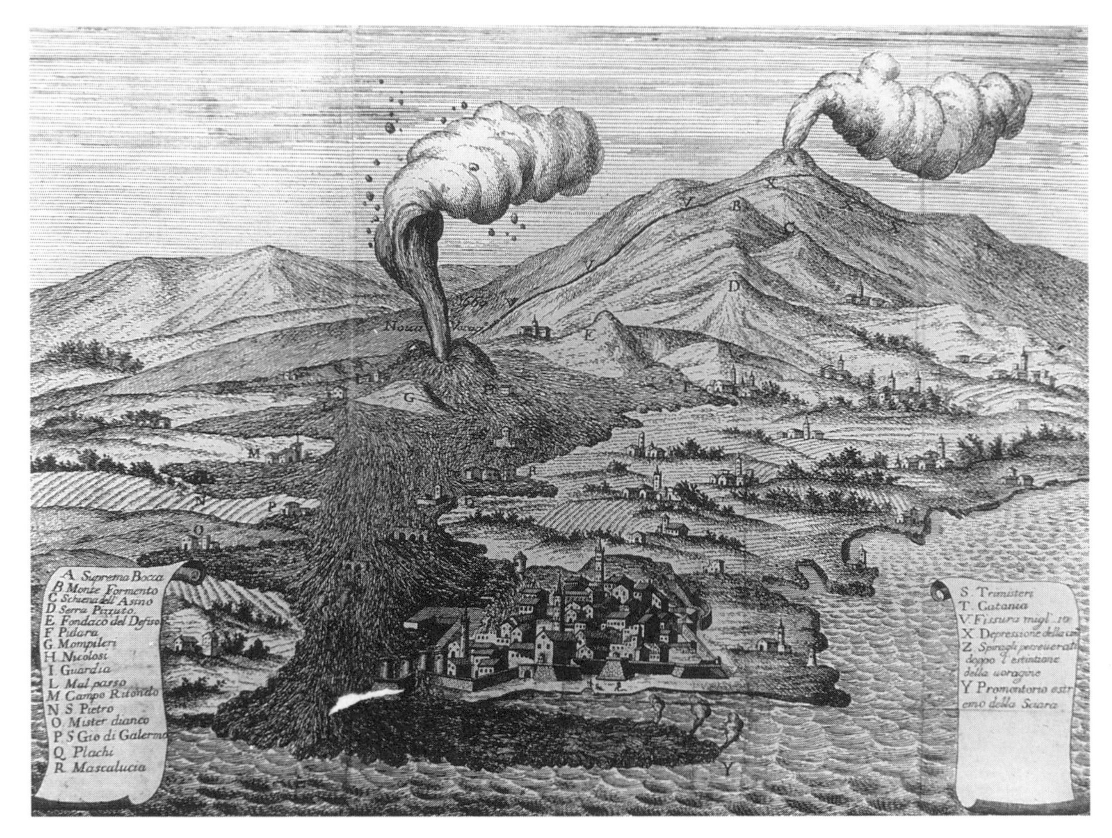

7.7. Lava flow from Mount Etna volcano enters the city of Catania in Sicily in 1669. Most of the lava issued from the new crater Monti Rossi, on the southeast flank of the volcano. Engraving (1792) by F. Anderloni, for Lazzaro Spallanzani's book *Viaggi alle due Sicilie e in alcune parti dell'Appennino*. From Pietro Bembo, *De Aetna* (Palermo: Sellerio editore, 1981).

> Once upon a time in Palermo, years ago, in company with some clever friends, I began to sublimate a quantity of sulfur, which altogether (in several vessels) did not reach the weight of fifteen pounds. The stench of this blessed sublimation was so irritating, not only to us who caused it, but to the whole vicinity, to all the streets around—I was just about to say to half of Palermo—that a huge outcry arose in the district, such an intolerable grumbling that we were obliged to hurry to the sudden remedy of extinguishing the fire and even after it was extinguished it left behind the infamy of its defect for two days, and led to accusation against that assembly of alchemists. Now...if a few pounds of sulfur was enough to pollute a large city, what should happen when not ten, not a hundred, not a thousand, not a million, but innumerable millions of millions of pounds of sulfur was burning for about a month in a favorable north-east wind? Should not the smell of it be carried to me so that I could enjoy it?

But d'Arezzo observed no such odor from the Etna lava, even when he was close enough to touch the "flowing fire," and concluded it could not be sulfur. He then offered his own revolutionary theory of the lava—that it was actually a glassy substance, the product of fusion in exactly the same manner as glass is made in a glassblower's furnace:

> Now, to tell you what I think, and to submit to your good judgement; I say and affirm that those so-called rivers, or that sea, that we like to call flowing or living fire, is nothing else but glass, or to speak

> more correctly rocks and other such stuff vitrified. Such a new concept will seem strange to you, but I cannot give it up when I consider all the properties, all the attributes, all the shapes of those rivers and that sea; I judge and see clearly that they are nothing but vitrified material.

To test its viscosity, d'Arezzo pushed a stick into the lava and pointed out that its adherence to other surfaces it came in contact with is very different from that of molten sulfur, which tends to coat and stick very readily to any surface due to its low viscosity. He also noted that the lava does not give off any smoke, and that it was much harder to melt in a crucible than sulfur and bitumen.

D'Arezzo communicated his findings to Borelli, who published them in 1670 in a book on the 1669 eruption of Mount Etna. Borelli followed d'Arezzo's ideas closely, but extended the discussion to the question of the origin of the molten rock:

> The most learned Francesco d'Arezzo, a nobleman of Syracuse, brought to my attention that sulfur and bitumen, ignited and liquified inside the furnaces of Etna, can in no way be changed and transformed into those huge masses of blackened rock that we call lava; but we should rather think that the sand and earth of Mount Etna is roasted and boiled by the extremely hot fire in its furnaces, and turned into liquid glass, and soon afterwards, meeting the air, acquires solidity and hardness.

Borelli further proposed that some alkaline flux was probably added to the volcanic furnace to facilitate the melting. In August 1671 Borelli himself made an expedition to Etna to satisfy his curiosity and even reached the crater. D'Arezzo was an important pioneer in that he was the first to recognize that lava is a silicate melt, but his critical observations were strangely ignored or overlooked by later workers.

Saint Gennaro Against the Volcano

One of the largest and deadliest eruptions in Europe occurred in 1631 when Vesuvius again erupted. Although this event did not arouse interest among scholars on the nature of volcanism, it did prompt a number of Neapolitan artists to capture the spectacle on canvas. Their work was, however, primarily stimulated by the alleged role of Saint Gennaro, the Neapolitan patron saint, in quelling the fires of the eruption, and these paintings offer a valuable view of various stages of the eruption as well as the religious response to the disaster. Before 1631, Vesuvius was clad in forests and vineyards and the green, pastoral slopes of the mountain were no more threatening to seventeenth-century Neapolitans than they were to the Romans in Pompeii and Herculaneum in A.D. 79. In this time of suppression of scientific thought, geologic knowledge and ideas about volcanic processes were very rudimentary. While scholars were aware of that great eruption from the writings of Pliny the Younger, Pompeii and Herculaneum were still undiscovered and erroneously believed to lie under the modern cities of Torre Annunziata and Torre del Greco. There had been reliable reports of Vesuvius erupting in 1347 and again perhaps in 1500, but the last major eruption of comparable magnitude to the 1631 event was in A.D. 472, when the ashfall extended as far east as Constantinople (Istanbul) in Turkey and south to Tripoli in Africa.

There are no fewer than 232 known contemporary written reports of the 1631 eruption, from which we can reconstruct the following sequence of events. In the evening of December 15, a strong earthquake, followed by many other shocks, was felt in Naples. Around six-thirty in the morning the next day, Vesuvius began to spew out fire and ash. By ten in the morning a dark, plinian eruption column was seen to rise above the volcano, like the trunk of a huge pine, that mushroomed into a dark cloud over the region, causing much fear and panic in Naples. The eruption cloud soon blocked out the sun, resulting in near darkness, and ash began to fall on the city, with a strong smell of sulfur likened to "the stinking fog of Hell." At this point terror was so extreme that the clergy adopted the age-old custom of bringing out the relics of Saint Gennaro, in a procession toward Vesuvius, to placate the mountain. Born in the third century, Gennaro acquired fame during the persecution of the Christians by the Romans. When cast before the lions in the amphitheater at Pozzuoli west of Naples, the godly Gennaro used his powers to calm the beasts and they refused to eat him. The Romans finally managed to take Gennaro's life, beheading him in the amphitheater-like crater of Solfatara. The relics—the martyr's head and a vial with his congealed blood—are kept in the Duomo in Naples and are the city's most sacred religious artifacts. Today, it is said that the saint's blood miraculously liquifies and bubbles in the glass ampulla twice a year: in May, commem-

7.8. During the 1631 eruption of Vesuvius, the citizens of Naples brought out the relics of Saint Gennaro to quell the fires, as depicted in a contemporary painting by Domenico Gargiulo. From Soprintendenza ai Beni Artistici e Storici, Naples.

orating his death, and in December, commemorating his intervention in the 1631 eruption. If the blood fails to liquify, it is considered a bad omen for the city, usually heralding a plague, war, earthquake, or, worse, a volcanic eruption. To many Neapolitans the history of Vesuvius is an eternal battle between the evil netherworld, represented by the volcano, and the good Saint Gennaro, but in this giant chess game the patron saint of the city is usually given the final move, according to legend.

The procession of clergy, which was accompanied by princes and lords and the Viceroy Count Monterrey, was to no avail. Explosions, earthquakes, and ashfall continued, sending the aristocracy in flight to the countryside and leaving the panic-striken public in the crowded churches, convinced that the Day of Judgment was at hand. During the evening the earthquake shocks were very severe, especially between seven and ten, but the wind had shifted to the west, carrying the ashfall away from Naples. The eruption continued, however, with devastating consequences for the districts around Vesuvius. Around two in the morning on December 17 a glowing ash cloud issued from the crater, flowing through the valley between Monte Somma (the ancient volcano) and the Vesuvius cone. This was probably the most lethal part of the eruption. Several mudflows or lahars flowed down the northern slopes of the Monte Somma, also causing much devastation around the town of Ottaviano. Around eleven in the morning a violent earthquake shook the region, followed by a discharge of large pyroclastic flows down the volcano's flanks, causing great destruction in the towns of Torre del Greco and Torre Annunziata, about 6 to 7 km (4 mi) from the crater. When refugees from the countryside arrived in Naples that night and the next day, they brought terrible tales of devastation—floods of hot sand and ash had flowed over their farms and towns, toppling houses and continuing out to sea. The hot flow "set alight everything, brushed aside trees, destroyed houses, burying buildings, farms, men, women, oxen, horses and every other animal until it plunged into the sea." The damage was estimated at 10 million gold coins and the death toll was more than 4000 people.

The course of events during the eruption can also be reconstructed by studies of the volcanic deposits. First produced was a high eruption column that generated pumice and ashfall. This column collapsed repeatedly, generating several glowing avalanches of ash and pumice that flowed out along the major valleys on the slopes of Vesuvius. The eruption cloud caused an atmospheric disturbance that generated torrential rain on the volcano, creating huge floods that devastated the east side of the volcano with mudflows. Finally, the eruption decapitated the volcano, resulting in reduction of the cone's height by about 500 m (1640 ft).

Naples was an important center of art in the seventeenth century and many artists living there doubtless witnessed the eruption. In fact, three paintings and one brush drawing depicting the event have survived. Although they were primarily executed as religious paintings, they provide us with valuable information about the eruption and contemporary view of volcanic activity.[13] These paintings, by Scipione Compagno and Domenico Gargiulo, can be placed chronologically, showing stages in the advance of the procession to the Maddalena Bridge on the southeastern outskirts of the city toward the volcano. The first view is Gargiulo's *Eruzione del Vesuvio* (Figure 7.8). Here the procession, with Cardinal Buoncompagni and

7.9. The aftermath of the 1631 eruption of Vesuvius is shown in this contemporary painting by Scipione Compagno. Kunsthistorisches Museum, Vienna.

Viceroy Monterrey accompanied by great numbers of clergy and bareheaded aristocracy bearing lit candles, circles the piazza, possibly near Porta Capuana. A white canopy shields the relics, and Saint Gennaro hovers above, surrounded by a band of angels riding on a cloud: this image is surely the birth of the iconography of Superman. With outstretched hands, the saint directs his powers toward Vesuvius in an attempt to bring the eruption under control. On the periphery of the procession are the poor, ragged masses of Naples and the peasant refugees from the Campanian countryside, some carrying the casualties of the eruption toward the religious relics. Meanwhile, Vesuvius thunders in the background, a dark gray eruption cloud expanding from its summit. The surrounding countryside is a bleak and devastated desert, coated in grey volcanic ash.[14] The painting clearly depicts the generation of a pyroclastic flow or surge cloud, expanding from the base of the eruption column and spreading laterally down the flanks of the volcano as a dark current of hot pumice and ash. Modern studies of the deposit, in fact, show that the lethal part of the eruption produced a pyroclastic flow or "sand flow," as depicted on later etchings of the event, in keeping with Gargiulo's work. In contrast were the erroneous views of volcanologists in the nineteenth and even early twentieth century that the eruption had devastated the region with mudflows, water floods, and a huge lava flow that supposedly traveled toward the sea in the region between Portici and Torre del Greco west of Vesuvius.[15]

The next "snapshot" of the eruption is Compagno's painting *Il Vesuvio*.[16] Here the procession enters the field of view from the extreme right, and has reached the Maddalena

Bridge, just east of the city walls, where the Sebeto River flows into the Bay of Naples. A white canopy protects the glass reliquary containing the saint's remains. On the far left the artist is perched on a high cliff, with a sketchpad on his knee. In the center is Vesuvius in furious eruption, its slopes covered in ash and dark grey clouds billowing about the summit. The volcano is the focus in this work, unlike Gargiulo's, where the procession takes center stage.

Compagno painted another view of the eruption (Figure 7.9) in which activity has decreased, although a plume of ash-laden clouds rises from the crater and lightning bolts flash through the eruption column.[17] The volcano's flanks are barren after the devastation by ash flows. Refugees are still rushing toward Naples over the Maddalena Bridge, where a great tumult is caused by a knight who charges his mount through the frightened crowd on his way toward the devastated areas. The procession is gone and again the center of attention is the erupting volcano. In this heroic view, Compagno was portraying the creative and destructive forces of nature, and the work approaches a true naturalistic landscape.[18]

In another view of the eruption, an ink drawing by the Neapolitan artist Filippo D'Angelo, the good work of Saint Gennaro is shown in much more graphic detail and verges on the comical (Figure 7.10). The saint is depicted seated in the clouds, instructing a band of angels who are quelling the eruption by pouring buckets of water on the smoldering volcano. In the foreground, the people of Naples flee in panic along the harborfront.[19]

Although the main eruption of 1631 ended on December 17, minor activity continued

7.10. The 1631 Vesuvius eruption as seen by the contemporary Neapolitan artist Filippo D'Angelo (called Filippo Napolitano), brought out Saint Gennaro and his helpers, serving as a fire brigade, but also causing panic among the citizens. Pen and brown ink, wash. Acc. no. 1993.274. Gift of Janos Scholz, The Pierpont Morgan Library, New York.

until January the following year. Then, in 1649, after a break of seventeen years, Vesuvius started up again and remained in an almost continuous state of eruption to the end of the seventeenth century. The eruptions during that period were very small events in comparison to the eruption of 1631, but they inspired many artists to incorporate the fiery volcano in their work or create solely volcanic landscapes. Among these is a brush drawing by Guercino (Giovanni Francesco Barbieri), *Landscape with Erupting Volcano*, undoubtedly inspired by Vesuvius (Figure 7.11).

It is remarkable that the cataclysmic eruption of Vesuvius in 1631 had little or no impact on the development of ideas regarding the origin of volcanoes or their inner workings. Contemporary philosophers made little mention of the event, nor did they discuss the unusual effects of the eruption on the surrounding area, where more than 3000 people perished. A notable exception is the Jesuit Giovanni Battista Mascolo (1582–1656) who published a book in 1633 recording his observations of the eruption, accompanied by very informative drawings of the volcano before and after the eruption —with Saint Gennaro hovering, of course, above the crater. It is, however, mainly through the work of the Neapolitan artists that we have been able to see a true representation of the process. In 1658 Vesuvius erupted again, and a contemporary scholar Salvatore De Renzi describes the event: "Saint Gennaro, glad that his good people had shown him such great devotion, was content to leave it at a scare

7.11. Landscape with a Volcano. Giovanni Francesco Barberi, called Il Guercino. Brush and brown wash on blue paper. Acc. no. 1975.36. Gift of Janos Scholz, The Pierpont Morgan Library, New York.

7.12. An eruption of Vesuvius, seen from Naples, with Maddalena Bridge in the foreground. Eighteenth-century engraving (author's collection).

and ordered Vesuvius to stop, and damage was limited to the few cultivations burned by the ashes." Belief in the powers of Saint Gennaro has remained strong in Naples to this day. A statue of the saint, with arm outstretched toward Vesuvius to stop the lava, was erected in 1777 at the Maddalena Bridge (Figure 7.12).

8

Burning Mountains and Cooling Stars

Earth by heat becomes fire and by cold returns into earth.

Isaac Newton, 1730

The most valuable knowledge given to the world is the method of acquiring new knowledge—that is, the scientific method. The creation of scientific methodology by the French mathematician and cosmologist René Descartes revolutionized the way in which we do science, and this method made possible the great discoveries of Isaac Newton on the laws of planetary motion and the force of gravity. Newton was also curious about the inner workings of the Earth, but his explanations for the causes of volcanism were a direct result of his secret experiments with alchemy. The attribution of volcanic activity and heat in the Earth to exothermic chemical reactions between iron and such sulfur-rich minerals as pyrites was widespread among scholars in the sixteenth and seventeenth century. This idea, as noted earlier, had its origins in the tinkering of the early alchemists with sulfur. Chemical reactions with sulfur caused some other substances to be combustible. The use of oil of vitriol (sulfuric acid) and other compounds derived from sulfur by Valerius Cordus (1515–44) made this element a key ingredient in the alchemist's art. The relation of sulfur to fire was, for example, addressed by the Venetian military engineer Vannoccio Biringuccio in his work, *Pyrotechnia* (1540). Biringuccio for a time held a monopoly on the production of saltpeter (sodium or potassium nitrate), which, together with sulfur, is a crucial component in gunpowder. Very early on alchemists discovered that combustion, or burning, required a component that is abundant in air, which the early chemists Robert Boyle, Robert Hooke, and Richard Lower referred to as *spiritus nitro-aereus*—it is clearly oxygen. The turning point from alchemy to chemistry occurred with the publication of a book by Robert Boyle (1627–91) in which sulfur features prominently, even in the title: *The Sceptical Chymist: or Chymico-Physical Doubts and Paradoxes, touching the Experiments whereby Vulgar Spagirists are wont to Endeavour the Evince their Salt, Sulphur and Mercury to be the True Principles of Things*. The ideas Boyle expressed in his book became the basis of chemistry and were later adopted and developed further by the great French chemist Antoine Lavoisier (1743–94).

Sulfur also featured prominently in the chemical theory of Johannes Joachim Becher, a physician of the Elector of Bavaria and a professor at Mentz. He ascribed many special properties to sulfur, and recognized the formation of sulfuric acid from sulfur, comparing the process to combustion.[1] Perhaps no chemist was more enthusiastic in his science than Becher. In the *Physica Subterranea* (Subterranean Physics) he speaks of chemists as a strange class of mortals, with an almost insane impulse to seek their pleasure among smoke and vapors, soot and flame, poisons, and poverty. "Yet among all these evils I seem to myself to live so sweetly that I may die if I would change places with the Persian king." In 1697 George Ernest Stahl (1660–1734), the royal physician at Berlin, elaborated on Becher's studies in *Zymotechnia Fundamentalis* (The Doctrine of Fermentation). He maintained that the process of forming sulfur from sulfuric acid was analogous to the addition of some combustible element, which he termed *phlogiston* (Greek for combustible). But in 1774 Lavoisier's discovery of oxygen completely overturned the phlogiston theory: he demonstrated that the heat given off during combustion was due to a chemical reaction with oxygen in the air.

Another pioneer chemist was the talented natural philosopher Robert Hooke (1635–1703), who began his career as Robert Boyle's laboratory assistant in Oxford. Hooke was one of the first to make a clear distinction between heat, fire, and flame, and, together with Boyle, to study the various phenomena of heat. He developed a "nitro-aerial" theory of combustion, in which thunder and lightning were likened to the flashing and explosion of gunpowder during the combustion of sulfur and niter. Hooke also thought that violent storms were caused by a reaction between sulfurous and nitrous particles in the air. Volcanic activity he attributed, rather vaguely, to "the general congregation of sulfurous, subterraneous vapors." He remarked on the greater frequency of volcanoes on islands and in coastal areas, ascribing this to:

> the saline quality of the sea water which may conduce to the producing of the subterraneous fermentation with the sulfurous minerals there placed. These fermentations subjacent to the sea, being brought to a head of ripeness, may take fire, and so have force enough to raise a sufficient quantity of the earth above it to make its way through the sea, and there make itself a vent. The foment or materials that serve to produce and effect conflagrations, eruptions or earthquakes, I conceive to be somewhat analogous to the materials of gunpowder.

Hooke also claimed that geologic activity was on the wane, and that subterranean fuel had been more plentiful in the past history of the Earth, when eruptions and earthquakes were apparently more severe: "That the subterraneous fuels do also waste and decay, is as evident from the extinction and ceasing of several vulcans that have heretofore raged." This concept of a finite supply of "subterranean fuels" was widely accepted and later endorsed even by the great British physicist William Kelvin in 1889. Hooke's *Lectures and Discourses of Earthquakes and Subterraneous Eruptions* (written in 1668; published posthumously in 1705) was in some respects a revolutionary work, presenting several geologic ideas that anticipated the geologic theory of James Hutton, published in 1785.[2]

Hooke's nitro-aerial theory was applied by such alchemists as L. de Capola in 1683 and Martin Lister in 1693 to a variety of phenomena; its influence may be seen in the ideas of Isaac Newton and others on the role of combustion in the generation of volcanic heat. Newton concerned himself with the process of volcanism, but his theories on the causes of this phenomenon were derived from contemporary alchemists and early chemists and, unlike his brilliant work on celestial mechanics, they have not stood the test of time. His views on volcanism or on the causes of "burning mountains" are clearly a direct outcome of his secret work on alchemy, in particular through his friendship with his older contemporary Robert Boyle. In his *Opticks*, Newton stressed the role of fire in the transmutation of bodies into light and also proposed that heating brought about the vibration of fundamental particles, or atoms, in matter. In his alchemical experiments, carried out in a wooden shed in the garden behind his rooms at Trinity College, Cambridge, he had repeatedly observed the evolution of heat, or exothermic reactions, when certain substances were mixed together, such as "when *aqua fortis*, or spirit of vitriol, poured upon filings of iron dissolves the filings with great heat and ebullition." With some other sulfurous mixtures, "the liquors grew so very hot on mixing as presently to send up a burning flame," and "the *pulvis fulminans*, composed of sulphur, niter, and salt of tartar, goes off with a more sudden and violent explosion than gunpowder." Private correspondance with Richard Bentley (1692) shows that Newton had already formulated some opinions about the origin of heat in the Earth and its intensity and had compared it to the irradiance received by the Earth from the Sun: "I consider that our earth is much more heated in its bowels below the upper crust by subterraneous fermentations of mineral bodies than by the sun." In an experiment described in *Opticks*, Newton tested his fermentation hypothesis and drew his conclusion on the causes of volcanism:

> And even the gross body of sulphur powdered, and with an equal weight of iron filings and a little water made into a paste, acts upon the iron, and in five or six hours grows too hot to be touched and emits a flame. And by these experiments compared with the great quantity of sulphur with which the earth abounds, and the warmth of the interior parts of the earth and hot springs and burning mountains, and with damps, mineral coruscations, earthquakes, hot suffocating exhalations, hurricanes, and spouts, we may learn that sulphureous steams abound in the bowels of the earth and ferment with minerals, and sometimes take fire with a sudden coruscation and explosion, and if pent up in subterraneous caverns burst the caverns with a great shaking of the earth as in springing of a mine.

The Earth as a Cooling Star

During the latter part of the seventeenth century, several philosophers adopted the view that volcanism was due to original or primordial heat in the Earth, and thus not related to combustion or exothermic chemical reactions. The first of these was the noted French mathematician and philosopher René Descartes, whose works were to have a profound influence on the evolution of thinking about the Earth as well as its origins. More than any other, Descartes challenged the contemporary mixture of fact and fancy that characterized natural philosophy in the seventeenth century and applied a new method of reasoning for seeking knowledge in any scientific field. He developed a new atomic view of a physical world consisting of invisible particles of matter in motion in the ether. Everything in nature could be explained by the mechanical interaction of these particles.[3]

Descartes' impact on geology stems largely from a theory presented in Part IV (The Earth) of his *Principa Philosophiae* (1644). The solar system originated, he proposed, as a series of "vortices," with the Earth as a star, "differing from the sun only in being smaller," and collecting dense and dark matter by gravitational attraction and losing energy as matter falls into place by gaseous condensation. He divided the Earth into three regions: a core consisting of incandescent matter, like that of the Sun, a middle region of opaque solid material (formerly liquid but now cooled) and an outermost region, the solid crust. All these layers, which he showed in a series of clear cross sections, had been arranged in this concentric fashion by virtue of their density. In this scheme, he maintained, there was definitely enough primordial heat remaining to supply any volcano—a radically different proposal from the common concept at the time of a heat source from subterranean combustion of coal, bitumen, and sulfur, or the deep-seated chemical reactions of the alchemists.

Descartes' ideas clearly influenced Gottfried Wilhelm, Baron von Leibniz (1646–1716), a German mathematician and philosopher who proposed in his work *Protogaea* (1693; published 1749) that the Earth must have existed originally in a state of fusion and had thus acquired its spherical form and concentric shell structure, with denser metals concentrated in the center. Lacking an independent source of heat, the planet had cooled by simple conduction over geologic time, forming a stony and irregular crust. This idea was supported "by the traces that still subsist of the primitive face of nature" in the form of scoria and volcanic glass produced by fusion.

Leibniz was not, however, prepared to entirely abandon the notion of combustible materials in volcanic activity, and wrote in regard to deposits of coal and sulfurous materials that "it is not unreasonable to believe that since the Deluge there have been partial fires, the date of which is not known, but which occurred at a time when combustible substances were more plentifully distributed in the thickness of the earth than they are now." Leibniz points out that even the Scripture hints that the Earth contains a fire that will reemerge. "For the greater part they believe, in effect (as the sacred writers, in their way, tend to imply), that there are raging fires at the center of the Earth from which flames can break forth in eruption."

In 1696 the British mathematician and theologian William Whiston published a *New Theory of the Earth*, in which he proposed that the Earth had been formed by the cooling of a comet, accounting for internal heat in a manner similar to Descartes. The French scholar Dortus de Mairan also argued in 1719 for a "central" heat in the Earth, and concluded, on the basis of surface temperature data from many parts of the globe, that there was a continual flow of heat from that center. Building on the vortices proposed by Descartes and Leibniz, the German philosopher Immanuel Kant (1724–1804) put forward in 1755 the nebular hypothesis, according to which the solar system had evolved from a rotating mass of incandescent gas. In 1796 this theory was further developed by the French mathematician Pierre Simon de Laplace (1749–1827), and is now known as the Kant-Laplace nebular hypothesis. It states that the Earth and other planets had coalesced from a mass of hot gas rotating around the Sun. In essence, this concept has survived to our day.

The influence of the Descartes' concept of a cooling star is also apparent in the work of the English clergyman and Earth theorist Thomas Burnet (1635–1715), who in 1681 published *Telluris Theoria Sacra* (The Sacred Theory of the Earth). This discourse on the original structure and history of the Earth was meant to be consistent with orthodox religion. According to Burnet, the Earth was an egg-like structure in which the central fire formed the yolk, surrounded originally by a huge liquid mass, "the great Abyss." Gradually, as the Earth became more stable, a smooth and level crust evolved, forming the antediluvian world, which was inhabited by early humanity in purity and innocence. Then came a universal catastrophe: heat generated in the liquid interior opened up crevices, created gigantic earthquakes, and split open the Earth's surface, transforming this terrestrial paradise into the irregular and wrinkled globe we know today, with the mountains of the Earth the ruins of the broken world. In Burnet's view, volcanoes are generated by the combustion of fossil oil substances and their destructive activity will end in the complete burning of the planet. This idea of the end of the world by fire, followed by a new Paradise, was compatible with religious dogma, as shown by quotations from *Origines* and Saint Augustine. Burnet also predicted a future geologic apocalypse: "How is it possible to confide in a transient world, which will be reduced to cinders and smoke in the space of a century or two?...It is reasonable to conclude that there are no more than fifteen hundred years to go till the end of the world."

Following Newton's great discovery, published in 1687, gravitational energy was regarded as by far the most important energy source in the physical universe. However, Newton was greatly interested in the idea of primordial heat in the Earth, and carried out some interesting and influential calculations on the rate of cooling of astronomical bodies: "A globe of red-hot iron equal to our Earth, that is, about 4,000,000 feet in diameter, would scarcely cool in an equal number of days, or in about 50,000 years." This hypothesis prompted the French scholar and pioneer of Earth science Comte George-Louis Leclerc de Buffon (1707–88) to carry out an experiment on the cooling rates of iron spheres to estimate the age of the Earth—which led him to the theologically dangerous conclusion that it had taken the Earth about 75,000 years to cool to its present temperature. This was, of course, radically different from the scripturally calculated age of the Earth's creation by Archbishop James Usher (1581–1656), who declared that the creation had occurred on October 26, 4004 B.C. at 9 A.M.

Born Georges-Louis Leclerc, de Buffon became a person of consequence in France, and his works *Théorie de la terre* (1749) and *Les Epoques de la nature* (1775, 1778) were to have a great influence on thinking about the Earth. In the latter work he attempted to synthesize Earth's history from the beginning. Following Descartes in considering the early Earth as a globe of molten matter that had solidified and cooled from the surface inward, de Buffon had the bright idea of calculating the planet's age by extrapolating the cooling rates of metallic spheres. Although his experimental methods were crude and his interpretations of the results questionable, de Buffon nevertheless ranks as one of the very first scholars to attempt to understand the Earth by experiment. At an iron foundry in Burgundy, de Buffon had his workmen cast ten balls of iron and nonmetallic substances of various sizes, up to 13 cm (5 in) in diameter. The balls were heated to near melting and then allowed to cool to air temperature while de Buffon sat, in the dark, in his thermally stable cellar laboratory, measuring the time it took for a white-hot ball to fade to invisibility. The cooling time, he observed, increased with the increasing diameter of the spheres; based on this observation he calculated that the cooling of a sphere of molten matter the size of the Earth would require only 2936 years to "consolidate to the center." He estimated that it took 25,000 years before water condensed on the surface of the cooling Earth, and 74,832 years before it cooled to its present condition. About 168,000 years after its formation, he concluded, the Earth will again be completely frozen and all life will have disappeared from its surface. Thus, de Buffon was the strongest proponent of the hypothesis that the Earth had began as a star, subsequently cooling from the outside. Later unpublished manuscripts show, however, that after he became aware of Newton's work, he felt that these figures were underestimates.[4] He then proposed 117,000 years, not 2936 years for the initial solidification of the globe—and that the Earth's age was nearly three million years. But he was reluctant to publish his revised estimates because he felt that such a great age of the Earth would seem inconceivable to his contemporaries. When his crude calculations on the cooling rates of the Earth were repeated by Lord Kelvin a century later, they yielded results that threw geology into total disarray, as discussed in Chapter 13.

Although de Buffon proposed that residual heat of the Earth was a legacy from its creation from the Sun, he nevertheless maintained that volcanoes derive much of their heat from the combustion of bitumen and sulfur within the Earth. He argued that during geologic time, vast quantities of vegetable and other organic matter were swept down into deeper parts of the Earth by geologic processes and these carbonaceous accumulations provided fuel for volcanic eruptions. He thought that volcanic energy arises from "the effervescence of the pyritous and combustible stones." However, volcanoes can only become active by "the conflict of a great mass of water with a great body of fire," and hence were always located near the sea. In his *Epoques de la nature* (1807), de Buffon speaks of the source of these combustible materials:

> These materials have been carried down and deposited in the low places and in the fissure of the rock of the globe where, finding substances already sublimated by the great heat of the earth, they formed the essential material for the alimentation of the volcanos to come. I say "to come," for there was no volcano in activity before the establishment of the waters. They could not begin activity, or at least a permanent activity, until after a lowering of the waters. One ought to distinguish terrestrial volcanos from marine volcanos; the latter make explosions, momentary so to speak, because at the instant the fire is started by the effervescence of the combustible and pyrite-bearing stones, it is immediately extinguished by the water which covers it. The land volcanos have, on the contrary, a lasting activity and one proportionate to the quantity of material they contain. These materials need a certain quantity of water to enter into effervescence, and it is only then, by the shock of a great volume of fire against a great volume of water, that they can produce their violent eruptions. It is for this reason that all the volcanos at present activity are in the islands or near the seacoast and that one can count a hundred times more extinguished ones than active ones; for, as the waters withdraw too far from the foot of the volcanos, their eruptions gradually diminish and finally cease.

De Buffon's dual view on the origin of heat in the Earth was without doubt influenced by the earlier writings of another French nobleman. In 1721 Benoit de Maillet (1656–1738), diplomat and traveler, completed the *Telliamed* (de Maillet spelled backward), a popular book proposing a theory of the Earth that greatly affected many of the leading naturalists of the time.[5] The book was not published until ten years after his death with the author's own name, but the unpublished manuscript had been circulated widely, causing violent controversy. As a consul of France in Italy, de Maillet came into contact with volcanism and, inspired by Italian scholars in this field, delved into the study of the Earth.

The *Telliamed* builds on Descartes' theory of a cooling Earth and considers the world as made in a diminishing sea, where sediments are laid down to form rocks. De Maillet assumed that all the heavenly bodies, including Earth, go through alternating luminous phases, when they behave as suns, and dark phases, when they behave as planets. The Earth, he maintained, was in the latter phase, and he drew an analogy between the Sun and the former condition of the Earth: "What has formerly happened, and still daily happens in the sun, tells me that it is a globe completely on fire, similar to ours, which, as yet, has barely reached such conditions and only in a few places; that its seas of fire are consuming its substance; that there have been times when these blazing seas have been covered with the scoria of the substances which serve as their fuel, as shown by the spots observed from time to time in the sun, which afterward disappear."

In stage four of his theory of the cyclical evolution of celestial bodies, he envisioned a cooling planet, with water condensing at its surface and life developing in the oceans. Although de Maillet was confident that internal heat in the Earth was inherited from the cooling star, he leaned toward a theory of internal combustion for the origin of volcanoes:

> Indeed, the volcanoes undoubtedly originate from the oils and fats of all the different bodies buried in the substance of these mountains. All the animals which live and die in the sea (and some of them are of enormous size, such as whales, from which we obtain such great amount of oil), so many rotten trees, decomposed plants and grass, all these are part of the deposits which the sea has built. It is

with these oleous and combustible substances that the mountains of Vesuvius and Etna, and some others, which like them vomit torrents of fire, have their deep-seated parts filled. Is not the coal found in England and in so many other countries an accumulation made by the sea, in the places where it occurs, of rotten weeds and of fish fat? Is it not the reason for its being combustible, as well as bad smelling? It is to these volcanoes, whether active or not, that we owe all minerals and metals, gold, silver, copper, lead, tin, iron, sulfur, alum, vitriol, and quicksilver, which their fire has at first deposited on the sides of the vents opened up by their flames, like the soot of the wood and coal that we burn is deposited on ours. The fat and oil of all the animals, fish, and other bodies, which may serve for the inflammation of opaque bodies [i.e., planets or cooled suns], are concentrated in certain places where, in time, they are set on fire. Such is the origin of volcanoes, which eventually communicate with each other, inflame the entire globe, deprive all its animals of the power of generation, and make a true sun of it.

De Maillet even predicted that eventually the Earth would lose its inhabitants due to the general burning of all its volcanoes and was particularly concerned about the "increasing number of our volcanoes, which are already so numerous in America. If ...the volcanoes increase in number and threaten the globe with generalized fire, men will at first try to escape death by moving as far away as possible from volcanoes, until the generalized fire reaches all parts of the Earth, a process which shall take many centuries in spite of the small size of our globe."

Most of these scholars had no firsthand experience of volcanic activity and based their ideas of volcanism entirely on written accounts. In the first half of the eighteenth century an Italian priest was, on the other hand, busy examining volcanic landforms and studying accounts of contemporary eruptions. The Venetian Abbe Anton Lazzaro Moro (1687–1764) was one of the main proponents of a fire at the center of the Earth; he considered mountain building and volcanic activity a primary consequence of the uprush of fiery gases from the Earth's interior. In 1707 an eruption within the caldera of Santorini led to the formation of the island of Nea Kameni, "thrust out of the sea mountain high" according to Moro (1740). He also cited the formation of the volcano Monte Nuovo in 1538, when it emerged near Pozzuoli in the Bay of Naples and grew over 150 m height in a few days. Moro, who had assembled records of sixteen islands that had been similarly built up by volcanic eruptions, was so impressed by the events that he proposed a similar mechanism to account for the origin of nearly all mountains. Violent subterranean fires had rent open the Earth in the past, discharging vast quantities of material, such as sand and stones, both solid and liquid, along with metals, sulfur, salts, bitumen, and many other mineral substances. Loose materials gradually piled up on the ocean floor around the new vents, and eventually the eruptive centers rose above the ocean surface, forming new islands. "Since the primitive mountains were born at the beginning of all things, when they were thrust out of the bosom of the Earth by subterranean fires and raised above the surface of the water that formerly covered all, they vomited various sorts of materials from their open mouths and caverns and these materials, either as running rivers or as rain falling from above, ran down and spread, one after the other, over the slopes of these mountains, just as we sometimes see it happen in Vesuvius, Etna, and in other similar flaming mountains."

On the origin of the subterranean fire he had a ready answer: "It pleased the great Creator of all things, when the dry land was to appear on the third day according to the sacred account in Genesis, that great subterranean fires should be kindled." With the birth of the Nea Kameni, Moro was convinced he had found the answer to resolve a number of geologic problems, including the occurrence of fossils in mountains:

> At daybreak on Monday 23 March 1707, there was observed in the bay of Santorini island, between the two Bracian islands, commonly called the Greater and Lesser Kameni, something that resembled a swimming rock, and that was at first believed to be a wrecked ship (Figure 8.1). Several sailors went in haste to identify the supposed ship, but they soon discovered to their immense surprise that it was a reef beginning to come up out of the depths of the sea. The next day several other persons attempted to land on the same reef, which was moving and moreover growing visibly, and they carried away several comestible things, among others some oysters of an extraordinary size and of exquisite flavor. Two days before the birth of this reef the entire island of Santorini was shaken by an earthquake soon after midday, which can only be attributed to the movement and detachment of the great mass of rock which the Author of nature had hidden from our eyes for so many centuries. This reef, without other disturbances, continued to grow until June 4, at which time it already occupied a space of nearly a half mile long and stood twenty-five feet over sea level.

But Moro was also concerned with the problem of space related to the movement of large masses of molten rock within the Earth.[6] Did his hypothesis imply the creation or existence

8.1. The underwater eruptions in the caldera of Santorini led to the emergence of new volcanic islands, as shown in this 1866 engraving (author's collection).

of great empty spaces inside the Earth? "Whatever space might have opened up by means of the blazing subterranean fires among those deep-lying materials...remains filled by the fiery igneous material until some opening or hole is made in the outer surface. When this happens and the fire has diminished, the space remains filled with air or with other volatile fluids. In this manner, every danger of vacuum is avoided. Second, if Empedocles' opinions should prove true, who imagined that within the rather thick hull of the Earth all of the internal cavity is full of fire, it is possible to say that the vacuum is avoided when each mountain or island is born, because when the portion of the Earth which that mountain or that island formation is raised and moved away from the center, another portion of earth in another part of the terraaqueous globe sinks toward the center." Moro then appealed to God's will in lighting the subterranean fires.

9

Columnar Basalt and the Neptunists

A volcano is not made on purpose to frighten superstitious people into fits of piety, nor to overwhelm devoted cities with destruction.

James Hutton, 1788

Some rocks, which were later found to be of volcanic origin, were thought by medieval philosophers to have originated by precipitation from water. Thus, the Swiss philosopher Konrad Gesner was convinced that the perfectly symmetrical and hexagonal basalt columns of many lava flows had precipitated as giant crystals from a primordial ocean (Figure 9.1). The oldest known pictorial representation of such a formation is found in his book on rocks, *De Rerum Fossilium* (1565). The basalt columns he illustrated were derived from a drawing by Kentmann, who was so struck with the strong resemblance of them to quartz crystals that he decorated them with the pyramidal terminations typical of that crystal.

Bernard Palissy, in *Discours Admirables*, also categorically stated that all crystal forms originate in water,[1] and more than two centuries later, from 1748 to 1753, the reverend Richard Pococke (later Bishop of Ossory) published three papers in the *Transactions* of the Royal Society of London, claiming that the columns of the Giant's Causeway—those huge six-sided pillars of basalt that rise from the sea in Northern Ireland—were formed by precipitation from an aqueous medium. Writing in 1776, Collini also considered these columns to be giant crystals. In 1722, Emanuel Swedenborg attributed Swedish basalts to a sedimentary origin—as did his

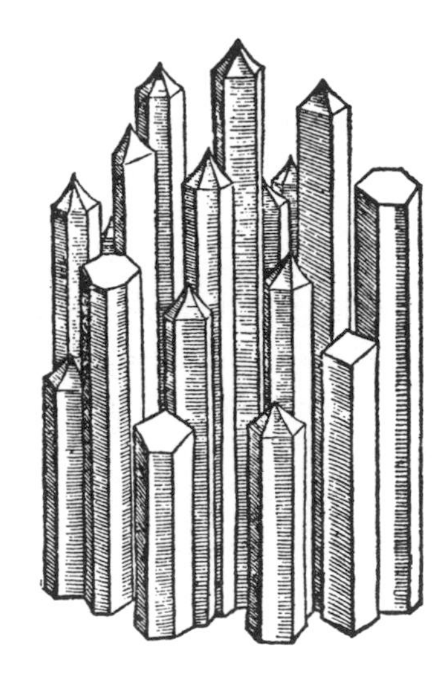

9.1. In 1565 Konrad Gesner viewed columnar basalt as giant crystals precipitated from a primordial ocean. It was later realized that these structures are solidified lava. From *De Rerum Fossilium*, Zurich.

famous countryman the systematist Carl von Linnaeus in 1756, after a study of the basaltic traps in Kinnekulle in Gothland. In *Observations of Natural Things*, Swedenborg was the first to refer to basalts as *traps*, a word derived from the Old Norse word *trappa*, or step in a stair, because of the characteristic structure of many columnar basalt formations. In 1756 A. Trembly also described basalt columns in western Germany as crystals. The concept of a crystallization form for these highly ordered rock structures is not surprising, considering their regular shape; the pillars of the Giant's Causeway resemble in every respect giant crystals that might have precipitated out of ocean waters. Similarly, the Swedish mineralogist Tobern Bergman (1735–84) had proposed that all rocks of the Earth's crust, including basalts, were chemical precipitates from a solution.

Later work was to show, however, that the columnar structure of basalts is the result of volume contraction on cooling and solidification of molten lava, just as the surface of a mud flat develops polygonal cracks as it dries out. Yet before this interpretation was generally accepted, the debate on volcanic versus aqueous origin of basalt was to become one of the fiercest controversies in the history of Earth science.

Basalt is by far the most abundant volcanic rock on Earth, with lavas on the ocean floor making up about 70% of the Earth's surface. It is found, of course, in most terrestrial volcanic settings and is the predominant rock type on the surface of the Moon. Basalt was mentioned by Pliny the Elder as early as the first century; in medieval times it is known from the writings of the mining engineer Georgius Agricola, who noted this dark, prismatic rock in several localities in Germany—one of them the rocky cliff on which the old castle of Stolpen in Saxony was built. In Germany basalt outcrops frequently take the form of horizontal capping on hills, and little about these outcrops suggested a volcanic origin. Another source of controversy involved vertical dikes of basalt, which later were correctly identified as solidified magmatic veins, or feeding arteries, of volcanic systems. In his book *The Hydrostaticks* (1672), George Sinclar describes basaltic dikes in the Scottish coalfields as *Gae of Whin-stone*, which had cut across coal seams, burning the surrounding coal. It is clear that Sinclar considered the

Gae or basaltic dikes a source of heat, although he did not specify their volcanic or magmatic origin.[2] The term *Gae* is probably related to the Old Norse *gjá*, meaning a fissure in the ground. Johann Gottlob Lehmann, professor of chemistry in Berlin and St. Petersburg, was one of the first to point out veins or dikes (*Ganggesteine*) that traverse stratified sedimentary deposits. Yet, like most others of his time, he proposed that the vein rocks were deposited from water. Lehmann was also one of the first geologists to carry out chemical experiments to test his theories, and he died in 1767 from injuries caused by an explosion of a retort filled with arsenic.

The basalt formation that drew the most interest from scholars in the seventeenth and eighteenth centuries was the Giant's Causeway in Northern Ireland (Figure 9.2). It displays on a colossal scale the columnar jointing typical of basalt, but its volcanic origin is not obvious—the region has been volcanically extinct for more than fifty million years. Scottish and Irish mythology attribute the formation to the giant Fionn MacComhal (Finn MacCoul), who built himself a bridge across the Irish Sea to attack his arch rival, the giant Fingal in the Scottish Hebrides who, according to legend, had made a similar bridge on the Isle of Staffa (Figure 9.3). Comments on the unique and peculiar structure of the Giant's Causeway date to 1693, but it is probably through numerous and widely circulated drawings and lithographs of this strange natural phenomenon that columnar basalt evoked such interest.[3] In Dublin, hearsay led scholars such as Sir Richard Bulkeley of Trinity College to publish a description

9.2. The columnar basalt formations in the Giant's Causeway in Northern Ireland played a key role in the debate between the Neptunists and Volcanists about the origin of basalt. Engraved by W. Deebie after a drawing by L. Clennell, 1813 (author's collection).

9.3. Columnar basalt formation on the Island of Staffa, Scotland. Eighteenth-century engraving (author's collection).

of the rock formation in the *Transactions* of the Royal Society of London in 1694, although he had never seen it. That same year the Reverend Samuel Foley, from firsthand observations, gave a fuller description of the phenomenon, accompanied by an engraving by Christopher Cole, the first known illustration of the Giant's Causeway.

The first to give a full account of the characteristic columnar jointing pattern of the Causeway was Doctor Thomas Molyneux (1697), who was also the first to point out that the rocks were basalts, or *Lapis Basaltes Misenus*. His article included another early engraved view of the Giant's Causeway by Edwin Sandys. Both illustrations were published in the *Transactions* of the Royal Society and received wide circulation among European scholars.

During increasingly more frequent voyages of scholars to other volcanic regions such as Iceland, Scotland, and central France in the period 1770 to 1790, the peculiar phenomenon of columnar basalt was commonly observed and thoroughly documented in lithographs and drawings by several other artists.

In 1740 the Irish artist Susanna Drury "made two very beautiful and correct paintings of the Giant's Causeway." Engraved and published in 1747, they were widely seen and quite influential in the progress of ideas on the origin of basalt rock.[4] The French pioneer volcanologist Nicholas Desmarest saw the engravings and in 1771 was the first to point out the volcanic origin of the rocks of the Giant's Causeway. Sir Joseph Banks, president of the Royal Society, described prismatic columnar basalts on the Scottish island of Cana in 1761 and on the Isle of Staffa in 1772. He remarked that the symmetrical basaltic columns of Fingal's Cave were "almost in the shape of those used in architecture."

Although Desmarest and others had already proposed that basalt rock was produced by volcanic eruption, many scholars held the contrary view, considering it a result of sedimentation from the ocean. By about 1790 the origin of fossils was no longer a subject of serious debate. These curious animal remains, found in rocks far from the ocean and on mountaintops, were recognized as residues of living creatures that inhabited the oceans of the past. Many saw their occurrence as evidence of Noah's flood, the great deluge that had wrought vast changes in the surface of the globe, as described in Genesis. This deposition of sediments from the universal ocean was the generally accepted explanation for the formation of the earliest rocks on the Earth. A group of eighteenth-century scholars, the Neptunists, held the extreme view, however, that *all* stratified or layered rocks had been precipitated chemically or mechanically from aqueous solution or suspension, including the layers of columnar basalt. The rocks they considered of sedimentary origin thus included many rock formations that were later shown to be lava flows and other products of volcanic eruptions.

The idea of precipitation of rocks from a great ocean had strong theological origins and derived from the legend of the Great Deluge. Perhaps the earliest philosopher to develop a geological cosmology involving the Great Deluge was Tomas Burnet who advanced the view of mountain landscapes as the ruins left by a great flood. Burnet's book *Telluris Theoria Sacra* (the full title reads: The sacred theory of the Earth containing an account of the Original of the Earth and all the General Changes which it hath already undergone or is to undergo till the Consummation of all Things; 1681) was entirely spun from his own vivid imagination, aided by a tour through the Alps. Burnet sought to include the mountains in the neatly ordered cosmology, as a part of the Creation and thus intended by the Almighty for some useful purpose. As mentioned in Chapter 8, his primordial Earth, a smooth and unblemished egg, consisted of a rigid shell and a vast interior reservoir of water. Eventually, the constant heat of the sun dried out and cracked the shell, releasing a vast flood from the deeps: the Great Deluge. These floodwaters brought about colossal change, carving landscapes on the face of the Earth and leaving behind mountains—"vast and undigested heaps of stone—the ruins of a broken world." The sediments laid down in the Great Deluge provided an irresistible starting point for the more empirical geologic evolution of the Earth of the Neptunists.

The leading figure of the Neptunist school was Abraham Gottlob Werner (1749–1817), born in Germany, in the mining region of Saxony. Werner was the first teacher of mineralogy and became the most popular and influential teacher and geologist of his time through his lectures at the Academy of Mines in Freiberg. Among his students were the famous geologists of his day, including Alexander von Humboldt, Leopold von Buch, Georges Cuvier, Johann Wolfgang Goethe, Jean François d'Aubuisson, and Robert Jameson. Of his teaching, Geikie has stated: "No teacher of geological science either before or since has approached Werner in the extent of his personal influence or in the breadth of his contemporary fame." Already in 1787 Werner was convinced that basaltic rocks were of aqueous origin—essentially the same view that had been put forward by both Konrad Gesner and Bernard Palissy two centuries earlier. Werner's studies of basalt at the summit of the Scheibenberg in the Erz Mountains further consolidated this opinion:

> The basalt rock is separated by several beds of sandstone, clay and greywacke from the basal gneiss. The transition from one stratified bed to the next in upward succession is quite gradual. Even the greywacke merges gradually into the clays below it and the basalt above. Therefore the basaltic, clayey and sandy rocks all belong to one formation, have all taken origin as moist deposits, precipitating during one particular epoch of submergence in this district. All basalt originally belonged to one widely extended and very thick layer, which has since been for the most part disturbed, only fragments of the original layer being left.

Werner's student Johann Karl Wilhelm Voigt (1752–1821) eventually opposed this interpretation, showing that the Scheibenberg basalt was an old lava of volcanic origin that had flowed over sediments, baking the underlying layers.

Werner considered the geological history of the Earth as taking place in five well-defined stages and, dismissing basalt as sedimentary in origin, saw the volcanic rocks such as pumice and scoria as merely a trivial component of his scheme:

1. Precipitation of sediments occurred from a global ocean to form the primitive rocks during the earliest period, giving rise to granite, gneiss, and porphyry.
2. During the second period, transition strata were formed by sedimentation from the ocean, creating slates, shales, and greywacke, which included fossilized fish.
3. Flötz, or secondary rocks, were formed during the third period, when ocean waters began to recede, depositing limestones, sandstone, chalk, and basalt as sediments.
4. During the fourth period, alluvial and derivative sediments were formed.
5. Finally, in the fifth phase, after the waters had disappeared from the continents, volcanic activity produced lavas, pumice, and tuffs.

His misattribution of basalt to an aqueous origin fundamentally shaped ideas about volcanism and the origin of the Earth. Yet the definition of basalt was to become the most bothersome problem of the Neptunists and led to the major geological controversy of the eighteenth century. Werner found basalt alternating with beds of sandstone and other sedimentary deposits of the ancient oceans and interpreted it as a chemical precipitate. But when he tried to differentiate between basalt and lava, his explanations became indefinite and obscure; in fact, he was trying to describe the difference between two identical rock formations. To overcome this obstacle, he explained that in some cases combustion resulted in the melting of basaltic rocks and when these liquids erupted they produced lava rocks with all the mineralogical characteristics of basalt (Figure 9.4). He also proposed that cross-cutting basalt veins or dikes were precipitates like mineral veins.

Werner's theory of precipitation of all rock formations from a universal ocean came to be known as the Neptunian Theory of the Earth. It had its attractions, conforming as it appeared to the teachings of the Book of Genesis. It also presented a view of a finished world, with a beginning and end. This concept was so flawed that it is difficult to see how geologists could ever have accepted it.[5] Yet Neptunist ideas were spread through Europe by the enthusiastic students of Werner, although little or not at all by his own writings. As pointed out by Sir

9.4. In October 1751 lava flowed from Vesuvius into the region of Boscotrecase, south of the volcano. For those who witnessed such eruptions and examined their lava products, it became difficult to accept the ideas of Werner and the other Neptunists that volcanism was a superficial phenomenon and that basalts were products of precipitation from the ocean. Eighteenth-century engraving (author's collection).

Archibald Geikie, "Fortunately for the progress of natural knowledge, Werner disliked the manual labour of penmanship. Consequently he wrote little." [6]

The Neptunist view was almost entirely developed by German geologists living in that part of Europe where there was no recent volcanic activity and even ancient volcanic rocks were rare. Werner thus considered volcanoes of minor importance, as accidental and relatively recent or postaqueous phenomena on the Earth. They owed their existence, he claimed, to the action of fire (much as the medieval philosophers had proposed), activated by the ignition of coal deposits or other flammable materials in the Earth: "Most if not all volcanoes arise from the combustion of underground seams of coal." To support this notion, he proposed that deposits of coal or other combustible materials were invariably present in the vicinity of volcanoes. It is truly amazing to find Seneca's combustion theory still invoked at this time, even among such learned men as Werner.

By the first half of the eighteenth century, chemists were familiar with the formation of crystals in the laboratory as a product of precipitation from an aqueous solution, but not from a high temperature fluid such as the silicate melt from which we make glass—and as represented by molten rock or magma in nature. Because Werner and his followers had recognized, quite correctly, that basalt was a crystalline rock, it was perhaps not unnatural that they also deduced its origin as a precipitate from an aqueous solution. Not until after the middle of the eighteenth century did the idea slowly begin to spread that crystallization could also occur as a result of the removal of heat from a silicate melt as well as from an aqueous solvent.[7] In 1761 Pierre Clement Grignon complained that chemists overemphasized aqueous systems, pointing out, after he managed to extract crystals from glass slags, that the products of glass furnaces compared to the products of volcanoes. The idea that magma was a solution that could produce crystals was beginning to emerge, and a group of geologists embraced the view that basalt owes its origin to solidification of magma or molten rock, either bought up to the surface during volcanic eruption or intruded into the Earth's crust. The Plutonists, as the adherents of this theory became known, found their intellectual leader in James Hutton in Scotland. A profound difference between the Neptunist and Plutonist theories related to the quantity and role of heat in the planet's interior. The Neptunists saw a negligible role for heat as a geological agent and considered volcanoes minor phenomena related to shallow-level processes—not as an indication of deep-seated heat. The Plutonists, on the other hand, regarded heat as the fundamental driving force for mountain uplift, folding, and deformation, pointing out the abundance of volcanoes, basalts, and granite as evidence of the melting of rocks at high temperature within the Earth. It is not my intention to review in depth here the course of this important debate, but to examine how the two schools of thought viewed the source of heat in the Earth and the nature of volcanic activity.

In Britain, the leading Neptunist was Richard Kirwan (1733–1812), an Irish chemist and mineralogist.[8] Attempting to relate the Earth's formation to the biblical accounts of creation, he interpreted the biblical passage "God said let there be light and there was light" as the bursting forth of volcanic eruptions on the Earth. To Kirwan, the Plutonist or Huttonian theory had one major flaw: "Where, then will he [Hutton] find those enormous masses of sulfur, coal or bitumen necessary to produce that immense heat necessary for the fusion of those vast mountains of stone now existing?"

The Scottish geologist Robert Jameson (1774–1854) also strongly opposed the Huttonian view of the Earth, particularly the volcanic theory of the origin of basalt and became, as noted, one of Werner's most ardent followers.[9] Jameson was a professor of natural history at the University of Edinburgh, but had deserted medicine for mineralogy in 1795 and by 1800 had gone to Freiberg to study the Neptunian system under Werner. In 1804 he returned to Scotland, an advocate of Neptunian geology. Apparently his lectures in Edinburgh on the topic were not always favorably received. One student in particular, Charles Darwin, resolved never to read another book on geology or study the science after hearing Jameson's uninspiring lectures: "I heard Professor Jameson, in a field lecture at Salisbury Crags, discoursing on a trap dyke, with amygdaloidal margins and the strata indurated on each side, with volcanic rocks all around us, and say that it was a fissure filled with sediment from above, adding with a sneer that there were men who maintained that it had been injected from beneath in a molten condition. When I think of that lecture I do not wonder that I determined never to attend to geology."[10]

The German explorer Alexander von Humboldt (1769–1859) also received a geological education with Werner at Freiburg and began his career as a Neptunist.[11] He was repulsed by the views of some of the Plutonists, which he considered unscientific and for good reason. Professor Samuel Witte of Rostock, for example, proposed that the pyramids of Egypt were ancient volcanoes because of their regular jointing pattern and peaked shape. In 1789 Witte published a treatise on the origin of pyramids in which he proclaimed that not only the Egyptian pyramids, but the Palace of Darius at Persepolis, the temples of the Incas in Peru, and the rock caves of Ellora in India were all the products of basalt eruptions. But Humboldt was to change his views on the nature of volcanic activity. Beginning in 1799 he undertook an expedition to Central and South America. While in Mexico, he made a thorough study of the deposits from the 1759 volcanic eruption of Jorullo, which had created a new mountain. On the basis of his work, he proposed that all volcanic centers are connected deep in the Earth. In his major work *Kosmos* (1844) Humboldt also proposed that petroleum was distilled by volcanic action from deeply buried sedimentary deposits (Figure 9.5).

The great German poet Johann Wolfgang Goethe (1749–1832) was another keen student of natural science who attended Werner's lectures and had thus been exposed to the Neptunian view.[12] However, in 1768 Goethe embarked on a journey to Italy that was to have a profound effect on his thinking. While visiting Catania at the foot of Etna, he encountered the lava that had destroyed much of the city in 1669: "Remembering what passions had been aroused before I left Germany by the dispute over the volcanic nature of

9.5. The summit of Cotopaxi volcano, Ecuador. When Alexander von Humboldt explored the many volcanoes of the Andes in 1799, he began to change his Neptunist views about the nature of volcanic activity. This engraving is from Humboldt's book, *Kosmos,* 1802 (author's collection).

basalt, I chipped off a piece; it is magma without any doubt. I did this in several other places so as to obtain a variety of specimens."[13] On reaching Naples, he was greeted by the sight of Vesuvius in full eruption, and a stream of molten lava flowing down its flank toward the ocean further demonstrated the role of volcanic action in the formation of basalt. Goethe's struggle with the Neptunist versus the Plutonist theory of the Earth may be manifest in his great drama *Faust*. It is in a dialogue between Seismos, personifying earthquakes and the violence of volcanic plutonic forces, and the Sphinxes, representing the calm and orderly progression of the Neptunian view of the origin of the world. Similarly, Mephistopheles represents the violent and destructive Plutonist view, while Faust reflects the dignified aspect of the Neptunist creation:

Mephistopheles

When God the Lord—wherefore I also know—
Banned us from air to darkness deep and central,
Where round and round, in fierce, intense glow
Eternal fired were whirled in Earth's hot entrail;
We found ourselves too much illuminated,
Yet crowded and uneasily situated.
The Devils all set up a coughing, sneezing:
At every vent without cessation wheezing:
With sulfur stench and acids Hell dilated,
And such enormous gas was thence created,
That very soon Earth's level, far extended,
Thick as it was, was heaved, and split and rended!
The thing is plain, not theories overcome it:
What formerly is bottom, now is summit.
Hereon they base the law there's no disputing,
To give the undermost the topmost footing:
For we escaped from fiery dungeons there
To overplus the lordship of the air,...

Faust

To me are mountain-masses grandly dumb:
I ask not Whence? and ask not Why? they come.
When nature in herself her being founded,
Complete and perfect when the globe she rounded,
Glad of the summits and the gorges deep,
Set rock to rock, and mountain steep to steep,
The hills with easy outlines downward moulded,
Till gently from their feet the vales unfolded!
They green and grow: with joy therein she ranges,
Requiring no insane, convulsive changes.

In the poem, a geological dialogue also takes place between the Vulcanist Anaxagoras and the Neptunist Thales:

Anaxagoras	Your mind is stiff and will not bow. What further argument is needed now?
Thale	The wave will bow to all the winds that play But from the rugged cliff it holds away.
Anaxagoras	It was volcanic gas produced this cliff.
Thales	In moisture is the genesis of life.
Anaxagoras	Have you, O Thales, ever in one night produced from slime a mountain of such height?
Thales	Never were nature and her living floods Confined to day and night and periods. Each form she fashions with due providence, Even in great things there's no violence.
Anaxagoras	But here there was! A fierce plutonic fire, Explosive gases of Aeolia, dire, Burst through that ancient crust of level ground That a new mountain might at once be found.

Here the Plutonist has the last word, although Goethe never fully abandoned the Neptunist view.

The French geologist Guy Tancréde de Dolomieu (1750–1801) also inclined toward the Neptunist view on the origin of basalts, even though he had traveled widely in the volcanic regions of Italy.[14] In Roman museums, Dolomieu studied many Egyptian basaltic objects, such as statues and sarcophagi, and declared that they were composed of a rock that showed no action of fire. Basalt was a well-known raw material for sculpture in Egypt in antiquity because it was favored for its extreme hardness, shiny luster, durability, and unusual dark green to almost black color. Similarly, the Italian geologist Abbé Spallanzani concluded that the Egyptian basalts were "produced by Nature in the humid way"—for example, by sedimentation. Although Dolomieu erred on the origin of basalts, he made groundbreaking observations on the petrography, or mineralogical and textural makeup, of other volcanic rocks. On his travels in Sicily and the Lipari Isles, he had paid particular attention to the study of volcanic rocks, including pyroclastic deposits and other loose ejecta, as well as lavas. He pointed out that gradations existed between the glassy rocks, such as obsidian, and coarsely crystalline lavas, in the grain size of the minerals. On this basis he proposed that many magmas do not represent a completely liquid state, but are viscous mixtures of crystals and liquid melt. He also concluded that magmas contain a combustible substance, which he thought was probably sulfur, that maintains the liquidity of the lava until the sulfur is completely consumed. In his view it was the expansion of this combustible substance that produced gas cavities and vesicles in volcanic products, such as scoria and pumice.

By the early nineteenth century, the Neptunist theory had become severely weakened and was encountering increasing opposition, especially when it was shown that the silicate minerals that compose crystalline rocks such as basalt and granite are insoluble in aqueous solutions at normal temperature.[15] Few adherents remained but John Murray (1802) attempted to rescue the theory from this trap by pointing out that silica is found in solution and precipitated from the high-temperature waters and exhalations of the geysers of Iceland. He also proposed that the primordial ocean was a very hot chemical soup that dissolved alkalis and silica, and was filled with the "saline, earthy and metallic matters."

In many cases converts from the Neptunist to the Plutonist view were made after scholars visited active volcanic regions. In 1803 one of Werner's students at the Mining School of Freiburg, Jean François D'Aubuisson de Vosins (1769–1819), published his study of basaltic rocks in Saxony (Germany), presenting a totally Neptunist interpretation. The following year he visited the volcanic region of Auvergne in central France and changed his views radically. When he saw the lava outcrops, merging downward into columnar basalt, "the facts which I saw, spoke too plainly to be mistaken. There can be no question that basalts of volcanic origin occur in Auvergne." From this he concluded that the basalts of Saxony must also be of volcanic origin.

Leopold von Buch (1774–1853) was probably Werner's most illustrious student and one of the most traveled field geologists. He left the Freiberg Academy after three years of study and went out into the world to put his teacher's theory of the aqueous origin of basalt to the test. In the face of the field evidence accumulated on his travels, von Buch also underwent a gradual conversion to the Vulcanist school.[16] In 1798 he visited the volcanic regions of Italy, where his Neptunian views received their first shock. In studying the deposits of ancient volcanoes near Rome and Naples for eight months, he encountered serious problems in trying to fit his observations to the Wernerian theory and complained, "Only two days at Vesuvius and all this confusion could be set right." But after he had wandered around Vesuvius and the Phlegrean Fields for eight weeks, he wrote: "My friends who believed that after seeing a real volcano, I now could say something definite on the diverging opinions about our basalts, will be badly disappointed." In 1799 von Buch studied the lavas of Vesuvius near Torre del Greco and could find little difference between those lava flows and basaltic rock. His Neptunist views were further weakened when he discovered that lava flows issuing from Vesuvius strongly resembled basalt. He was also impressed by the great manifestation of volcanic energy of Vesuvius, but looked in vain for evidence of the combustible coal or pyrite deposits in the region that could fuel the immense subterranean fire, as Werner had proposed. Yet, rather than abandon his Neptunist prejudice, he proposed that the basaltic lava represented a preexisting basalt of aqueous origin that had been fused deep in the Earth's crust under the volcano and then ejected as lava.

In 1802 von Buch also visited the Auvergne district for six weeks and he examined the rocks that Desmarest had attributed to a volcanic origin forty years earlier. He agreed that they were volcanic lavas, but still argued that they could have been a product of local melting and eruption of basalts of aqueous origin. In 1805 he was again in Naples with Alexander von Humboldt and the French chemist Gay Lussac and witnessed Vesuvius in eruption. The further he probed, the clearer it became that volcanoes were not local and unimportant factors in the history of the Earth, but of widespread and far-reaching importance in the formation of the Earth's crust, and that they had their origin deep within the Earth's interior. It was perhaps von Buch's travels in Scotland and his studies of the columnar basalt formations at the island of Staffa and the Giant's Causeway that finally led him to renounce the Wernerian theory on the origin of basalt. His work also demonstrated that basalt and granite were not merely precipitates from a universal ocean, but of igneous origin and formed by the crystallization or solidification of magma or molten rock.

Eventually von Buch proposed a volcanic origin of oceanic islands from his studies in the Canary Islands in 1815. There he devoted much energy to observing the form of volcanic mountains and developed his "Craters of Elevation" theory, which stated that volcanic mountains were thrust up by upheaval of the Earth's crust. Volcanoes, he noted, were generally conical in form and their strata dipped in all directions away from the central crater at the summit. Curiously, he did not recognize that this was due to the pouring out of lava and ash from the summit, but attributed their form to uplift, accounting for the craters through the bursting and collapse of a bubble-like mass at the summit. To support his theory, he proposed cavernous openings beneath the volcanoes, separated by walls or arches, and believed that these caverns were periodically subject to influx of great quantities of gases and lava. Although von Buch abandoned the Neptunist theory, he never abandoned the Craters of Elevation theory.

10

The First Field Volcanologists

Accurate and faithful observations on the operations of nature are not met with often. Those who have wrote most, have seldom been themselves the observers.

Sir William Hamilton, 1776

New perspectives toward Nature in the mid-eighteenth century led to increased familiarity with the Earth, and as travelers ranged more widely, interest in volcanoes and volcanic activity grew rapidly. Attention often focused on Vesuvius—with its frequent eruptions in the latter part of the century, it provided major displays of volcanic action, much to the delight of the cosmopolitan north European travelers on the Grand Tour. Other volcanic phenomena also became popular, requiring stops on travelers' tours, including volcanoes in the Auvergne region in France, Mount Etna in Sicily, and the columnar formations of the Giant's Causeway in Ireland and on the Scottish island of Staffa. Some travelers, among the rich and cultured of the Enlightenment, perhaps viewed themselves as the new philosophers of the world—admiring Nature for its grandeur and emphasizing its permanence, activity and creativity—but often had contempt for the narrow views of Christian religion.[1] Christian natural theology had deliberately minimized the role of volcanoes in Earth history because of their association with destruction; their only role was in the iconography of Hell. To enlightened scholars, on the other hand, volcanic activity must play a necessary and constructive role in the Earth's economy. The British scientist William Hamilton, for example, regarded volcanoes as an integral part of Nature, and his writings are saturated with contempt for Catholic superstition and for all those who thought of eruptions as a Hell-fire judgment.

Meticulous observations and beautiful pictorial representations in Hamilton's works greatly helped to win credence among scientists that volcanoes were important and positive Earth forces, not freaks of Nature: "Volcanoes should be considered in a creative rather than a destructive light" Hamilton wrote in 1786. After all, the effects of Vesuvius and Etna had been to enlarge the coasts of Italy and Sicily, not to throw the area into confusion. The work of Hamilton and the other Volcanists was of threefold importance. It focused attention on active forces of Nature that were possibly the products of a global central heat. It showed that such forces continue to this day—important in underlining the continuity of the Earth's present with its past and thereby encouraging actualistic methods. And it encouraged an extensive search for extinct volcanoes, identified by the presence of lava flows, basalt, or such landscape characteristics as conical hills with flat tops. The revolutionary finds of ancient volcanoes in France caused much attention, but it was also Hamilton's meticulous reports of an erupting volcano that helped establish volcanic activity as a major force in geology.

William Hamilton (1730–1803) was born into the British aristocracy, the fourth son of a Scottish lord (Figure 10.1). In 1761 he became a Member of Parliament, and three years

10.1. During his stay as a British diplomat in Naples at the end of the eighteenth century, Sir William Hamilton had ample opportunity to observe the activity of Vesuvius and became fascinated by volcanic processes. Eighteenth-century engraving (author's collection).

later was appointed Plenipotentiary and British Envoy Extraordinary to the court of Naples, where he lived until 1800. His extended stay in the city at the foot of Vesuvius led to a lifelong interest in the study of volcanic activity, and he may rightfully be called "the modern Pliny."[2] Hamilton published his findings on Vesuvius, Etna, and other Italian volcanoes regularly in the *Transactions* of the Royal Society of London, but his principal work was *Campi Phlegraei*, published in 1776. In this masterpiece he presented detailed observations of the volcanoes of Italy and even hired the noted artist Pietro Fabris to illustrate his findings. His interests also extended to archeology and he collected and published on the antiquities found in the excavations of Pompeii and Herculaneum. He sold his collection to the British Museum in 1772 and it became the nucleus of the Greek and Roman holdings. He continued to contribute to the museum, including a colossal head of Herakles found in the volcanic deposits at the foot of Vesuvius.

Hamilton's study of geology in Italy was thorough and energetic. He personally conducted fieldwork, explored valleys and ravines for volcanic rock formations, climbed volcanoes—in the early stage of his studies, he climbed Vesuvius no fewer than twenty two times in only four years—and made detailed observations of eruptions at close range, often in much personal danger (Figure 10.2). His physical condition must have aided in his field studies; Hamilton was described as a man "of spare figure and of great muscular power and energy." Unlike most of his contemporaries in the field, Hamilton was more concerned with observations than with the formation of hypotheses or theories about the Earth. In *Campi Phlegraei* he wrote:

10.2. The crater of Mount Vesuvius before the great eruption of 1767. When William Hamilton arrived in Naples, he was greeted by a major eruption of the volcano, which destroyed the summit cone. Following this, the volcano gradually built up another cone within the great crater. Engraving dated 1813 (author's collection).

> I do not wonder that so little progress has been made in the improvement of Natural History, and particularly in that branch of it which regards the Theory of the Earth. Nature acts slowly, and it is difficult to catch her in the act. Those who have made this subject their study, have without scruple, undertaken at once, to write the natural History of a whole province, or of an entire continent; not reflecting, that the longest life of man scarcely affords him time to give a perfect one of the smallest insect. It is to be lamented, that those who have wrote most, on the subject of natural History, have seldom been themselves the observers, and have too readily taken for granted systems which other ingenious and learned men, have perhaps formed in their closets, with as little foundation of self experience; the more such systems may have been treated with ingenuity the more have they served to mislead, and heap error upon error.

Hamilton thus fully grasped the perspective of time in geological processes when he wrote: "Nature is ever active, but her actions are, in general, carried on so very slowly, as scarcely to be perceived by mortal eye, or recorded in the very short space of what we call history, let it be ever so ancient." In commenting further on theorists who made speculations based on scanty observations he wrote: "There have been Naturalists, of such a wonderful penetrating genius, as to have thought themselves sufficiently qualified to account for every Phenomenon of Vesuvius, after having literally speaking, given the volcano, *un coup d'oeil*." Hamilton's meticulous observations and documentation of the growth of the cone of Vesuvius are a good example of his quantitative approach (Figure 10.3). Unlike many of his contemporaries, he also regarded volcanism as an important part of nature's progress, "an operation much less out of her common course than is generally imagined." He realized that progress in volcanology could be made by analyzing the chemistry and mineralogy of volcanic rocks, and sent many thousands of specimens back to England for analysis. Yet, the study of these specimens appears to have been fruitless because Hamilton never refers to the chemical results in his writings. By 1776 he had independently discovered the volcanic origin of basalt, two years after Desmarest had arrived at the same conclusion in the Auvergne district of France, as described below. Hamilton wrote: "Every day seems to produce fresh discoveries of ancient volcanoes, and it can now be no longer doubted, but that wherever Basaltic columns of the nature of the Giants Causeway in Ireland are found, there have Volcanoes existed, for they are merely Lava."

The seat of volcanic fires, Hamilton maintained, was deep and far from superficial, as had been previously thought by those who proposed combustion of sulfur or bitumen in the Earth's crust. He reasoned that fuel for the fire would be insufficient to support frequent eruptions if it were merely superficial, and the hollow beneath a volcano at shallow depth would cause it to collapse. In earlier work (1772), however, Hamilton had supported medieval beliefs of volcanic heat when he discussed the role of combustion of bitumen in supplying volcanic heat. In describing Solfatara volcano in Italy, he wrote:

> Another hollow...shews that the mountain is composed of strata of erupted matter, among which are large masses of bitumen, in which its former state of fluidity is very visible. Here it was that I discovered that pumice stone is produced from bitumen, which I believe has not yet been remarked. Some

> specimens show evidently the gradual process from bitumen to pumice: and you will observe that the crystalline vitrifications, that are visible in the bitumen, suffer no alteration, but remain in the same state in the perfect pumice as in the bitumen.

It is clear that Hamilton mistakenly identified the black volcanic glass or obsidian as bitumen in this instance.

In writing about the great 1783 earthquakes in Calabria, Hamilton attributed them to the same forces that gave rise to volcanism, and allowed himself to speculate on the fundamental source: "The destruction I have been describing may have proceeded simply from the exhalations of confined vapours, generated by the fermentation of such minerals, as produce volcanoes, which have escaped where they met with the least resistance, and must naturally in a greater degree have effected the plain than the high and more solid grounds around it." The generation of heat and vapors by fermentation of minerals recalls Newton's chemical theories, whose works Hamilton must have known.

Hamilton was one of the first students of volcanic rocks to differentiate between the rock types of tufa and lava. Tufa (tuff in modern usage) is a hardened deposit of volcanic ash and pumice, the products of explosive eruptions. Tufa rock is plentiful in Italy and commonly used

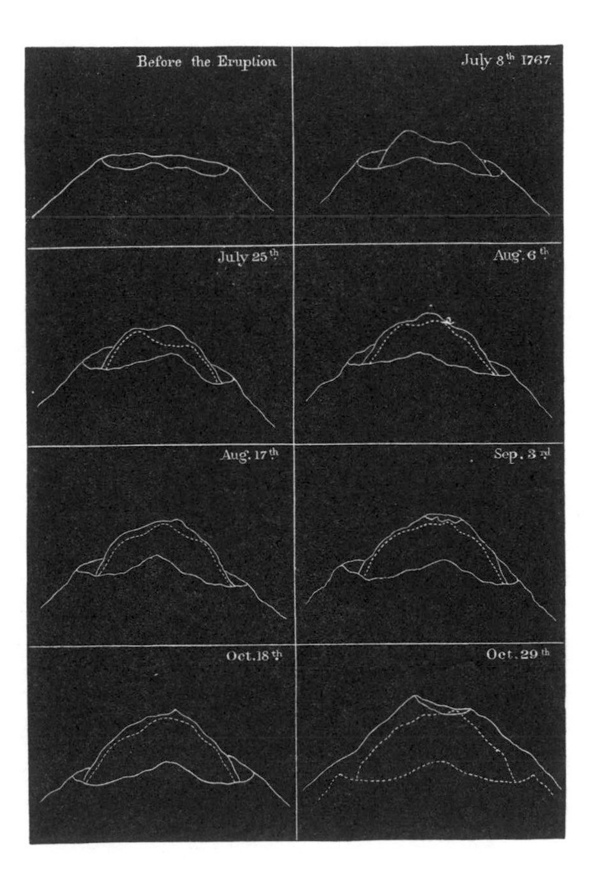

10.3. The outlines of the summit of Vesuvius volcano during the eruption of 1767, as drawn by William Hamilton at successive dates. From J. W. Judd, *Volcanoes* (New York: D. Appleton & Co. 1881).

as a building stone. Hamilton must have seen tufa in the walls of many Neapolitan buildings and probably visited the numerous tufa quarries around Naples and in the Phlegraean Fields. The rock, he believed, was produced when water came in contact with magma in volcanic eruptions and he proposed that the tufa material issued out of the volcano as a sort of cement:

> Tufa...seems to me to be, that fine burnt material which is called pozzolene, whose binding quality and utility by way of cement are mentioned by Vitruvius, and which is to be met with only in countries that have been subject to subterraneous fires. It is, I believe, a sort of lime prepared by nature. This, mixed with water, great or small pumice stones, fragments of lava, and burnt matter, may naturally be supposed to harden into a stone of this kind...Naturalists might adopt the Neapolitan term Tufa for that sort of mixture of ashes and pumice stones, which is thrown up by volcanoes in the state of liquid mud, and taken the consistency afterwards of a soft and light stone.

It is clear from Hamilton's discussion that he distinguished between relatively quiet eruptions resulting in the outpouring of molten lava and violent explosive eruptions that produced tufa and ashes. He correctly attributed this violent action to water coming in contact with hot magma within the volcano. In the absence of water, the eruption would be quiet, producing a lava flow.

During his excursions in the mountains near Naples, Hamilton noted that numerous veins or vertical sheets of basalt cut across older rock formations, such as in the cliffs of the ancient Monte Somma volcano that cradle the young cone of Vesuvius. He recognized these veins as dikes that had served as the feeder conduits of magma from depth to surface during earlier eruptions, and thus was one of the first volcanologists to ponder the plumbing system of a volcano.

Hamilton was also one of the first to recognize that lava flows can be differentiated from dikes and other intrusive igneous rocks on the basis of the bubbly, frothy, or vesicular nature of the margins of lavas. In *Campi Phlegraei*, he discussed the vesicular or porous upper and lower margins of lava flows "which must undoubtedly proceed from the impression of the air upon the vitrified matter whilst in fusion." By 1779 he had revised this opinion, concluding that: "When the pores of the fresh solid lava were large and filled with pure vitrified matter, we found the matter sometimes blown into bubbles in its surface I suppose by the air, which had been forced out at the time the lava contracted itself in cooling." Thus, he correctly deduced that vesicularity was the result of the exsolution and expansion of volatiles from the magma.

Hamilton's work on the active Italian volcanoes was read widely. For example, John Whitehurst (1713–88) was influenced by Hamilton's writings on Vesuvius and other volcanoes, and in his *Inquiry* (1778) Whitehurst associated volcanic activity with a central fire in the Earth and showed that it had earlier played a role in the geology of England. While working in the Derbyshire mining region of England, he made a crucial discovery when he recognized that massive rock layers, known locally as "toadstone," were in fact of volcanic or igneous origin. Other scholars were also captivated by the volcanic activity of Vesuvius at this time. Among them was Johann Ferber, a Swedish mineralogist, who made insightful descriptions of the volcanic rocks of Vesuvius and other volcanoes of the Campania. In Naples in 1755 Father della Torre, the Royal Librarian, compiled a complete record of all the eruptions of Vesuvius from 79 A.D. to his day.

Hamilton's fond interest in geology was shared by Frederick Hervey, Bishop of Derry and fourth Earl of Bristol (1730–1803). Spurred by accounts of the eruption of Vesuvius in 1766, and by his own observations at the Giant's Causeway and travels on the continent in 1770 to 1772, Hervey devoted much effort to exploring rock formations in France and Italy resembling those of the Giant's Causeway.[3] In 1773 he wrote:

> You have doubtless heard much of our Giant's Causeway: till lately it has been reckon'd single of its kind, but I have lately discover'd such varieties of the same sort both in France and Italy, and accompanied with such peculiarities of soil as can no longer leave the origin of this strange phenomenon a Problem: the entire district of Velay in France is compos'd of it; the villages, the Castles & the farm houses are built of these materials which are true Basaltine Stone, & every Isolated & Conical hill is compos'd of it; at the back of Clermont in Auvergne, a country strongly convuls'd and almost shatter'd by volcanos in the time of Sidonius Apollinaris Bishop of Clermont, Anno d: 450. There is among many other isolated mountains full of craters and covered to this day with cinders, lava and pumice-stone, one in particular compos'd entirely of these Polygon Basaltes which in such a situation leaves their origin unquestionable—to confirm this one need but observe the shooting of salts in any Chymical experiments or the formation of Sugar-candy in a sugar-house, to be convinc'd that these columns have likewise been in a state of Fusion, & owe their figure to the action of Fire—all this country of Ireland has in some remote age been equally agitated by subterranean fires and the numerous Sugar loaf'd & Isolated hills, full of Lava & Pumice-stone with Chrystal & Vitrifications, are strong proof of it—but when I see many of the same hills as full of marine bodies partly calcin'd & partly uncalcin'd mix'd with the same Volcanic matter, I cannot hesitate in supposing the whole mass thrown up like the Islands near Santorini in the Archipelago at the beginning of this century [1707], like Thera & Therasia in the same Sea at the time of Pliny, & like Delos & Rhodes itself long before his time, by fire under the sea.

Guettard and Desmarest

Most of the early ideas about heat in the Earth were based on what might be called "armchair geology," speculation derived from on little or no field observation. In the eighteenth century, the approach to the study of the origin of the Earth gradually evolved from pure speculation to a search for answers in rock formations exposed at the surface and in mine shafts. The labors of certain men, including those connected with the great mining industry in southern Germany, brought about a revolution in the study of the Earth and the invention of field geology. This philosophy was advocated as early as 1571 by the Danish scholar Peter Severinus, who advised his students: "Go my sons, sell your lands, your houses, your garments and your jewelry; burn up your books. On the other hand, buy stout shoes, climb the mountains, search the valleys, the deserts, the sea shores, and the deep recesses of the Earth. Look for the various kinds of minerals, note their characters and mark their origin. Be not ashamed to learn by heart the astronomy and terrestrial philosophy of the peasantry. Lastly, buy coal, build furnaces, observe and experiment without ceasing, for in this way and in no other will you arrive at a knowledge of the nature and properties of things." The philosophy advocated by Severinus for a new experimental science based on fresh observation was probably influenced by the example set by the great alchemist Paracelsus.[4]

Sir William Hamilton did indeed make important field observations, but they were on active volcanoes, where rocks were easily identified as the product of eruptions. The next great leap in volcanology occurred when geologists were able to identify ancient rocks as being of volcanic origin in regions far removed from active volcanism. This breakthrough took place in the middle of the eighteenth century in France, where two men were dedicating themselves to close observation and accumulation of facts about the Earth.

Jean-Etienne Guettard (1715–86) ranks with the founders of geology. Born into a pharmacist's family, Guettard developed a passion for natural history. He first studied botany in Paris, where he came into contact with the Jussieu brothers of the Jardin des Plantes in Paris, the leading botanists of the time. Later he became a doctor of medicine and was employed by the Duke of Orleans, whom he accompanied on travels throughout France. He also took charge of the Duke's important rock and mineral collection—which piqued his interest in geology. When Orleans died, Guettard received a pension and rooms at the Palais Royal in Paris. Now independent, he was free to continue his research in botany and geology.[5] His written contributions on natural history are astonishing in their number; he produced more than 200 papers, and a half-dozen volumes of observation. By 1734 he had been honored with election to the French Academy.

Guettard's influence on ideas concerning volcanism in particular was profound, and his discovery in 1751 of extinct volcanoes and their products in the Auvergne district "in the very sunlit heart of France" was an important step in the science of volcanology. We now know that these volcanoes were active about 7600 years ago. Guettard's discovery is worth recounting in some detail because it marks a turning point in the study of volcanoes and the birth of volcanology.[6] In 1751 Guettard and his friend Chretien-Guillaume de Malesherbes made a journey south from Paris to take the waters of Vichy. En route they made an excursion through the Auvergne region and noticed some very unusual black porous stones in mileposts along the Allier road, in pavements, and in the walls and buildings in the village of Moulins. These strange rocks greatly aroused their curiosity and Guettard immediately suspected that they were lava rock. (Earlier the two friends had visited Vesuvius and were quite familiar with the appearance of young volcanic rocks.) The pavements in Auvergne, made of short hexagonal columns, resembled the ancient pavements of roads near Rome and Naples that were composed of polygonal slabs of lava.[7] The rocks, they were told, were quarried near Volvic, north of Clermont. Guettard is then said to have exclaimed, "*Volvic, volcani vicus*," "volcano village," suggesting that the village had drawn its name in Roman times from the volcanic nature of the region. Could the Romans have been the first to recognize the volcanic nature of this part of France, a flourishing colony at the time of Nero? Guettard and

Malesherbes continued on to Vichy, where they found that the building at the famous water spring was also built of lava stone from Volvic. They then visited the Volvic quarry, where Guettard noted dipping layers of the rock, with scoriaceous or frothy upper and lower surfaces, as well as other unmistakable signs of volcanic activity; he concluded that the stones were indeed from a large lava flow (Figure 10.4). They were able to locate seventeen craters and volcanic cones in the area, now long peaceful and covered in woods. When they came to Puy de Dome, Guettard recognized that basaltic rock of the hardened lava had flowed from an ancient crater in the cone-shaped hill above.

In 1752 Guettard presented the results of his geological investigations to the Royal Academy of Sciences in a paper entitled "*Sur quelques montagnes de la France qui ont été volcans*." (On some volcanic mountains in France.) It may be said that the Auvergne district became the "graveyard" of Neptunism once his observations of ancient volcanic activity in central France were widely known.[8]

Guettard adhered to the conventional view (following his countryman de Buffon) of attributing volcanic eruptions to the combustion of flammable materials: "For the production of volcanoes, it is enough that there should be within these mountains substances that can burn, such as petroleum, coal or bitumen, and that from some cause these materials should take fire. Thereupon the mountain will become a furnace, and the fire, raging furiously within, will be able to melt and vitrify the most intractable substances."

Curiously, Guettard failed to place columnar basalt in its proper volcanic setting. He

10.4. Columnar basalt formation in the Auvergne district of France where Etienne Guettard made his fundamental discovery that ancient rocks in central France were of volcanic origin (author's collection).

noted such columnar formations in association with the lava flows he studied, but denied that they were related. "If a columnar basalt," he wrote, "can be produced by a volcano, why do we not find it among the recent eruptions of Vesuvius and other active volcanoes?" This was a fair argument, given the state of observations of lavas from active volcanoes at the time. Consequently, he concluded in 1770 that "basalt is a species of vitrifiable rock, formed by crystallization in an aqueous fluid, and that there is no reason to regard it as due to igneous fusion." Guettard may thus be considered the parent of both the Volcanist and Neptunist schools. Although his observations in Auvergne led to the Volcanist study of volcanic rocks in Europe, his proposal of an aqueous origin for basalt was to become a cornerstone of the Neptunist theory.[9]

About the same time that Guettard was making his discoveries in central France, Giovanni Arduino (1713–95), an Italian pioneer in geology, discovered that some ancient basaltic or trap rocks in northern Italy, which were interstratified with sedimentary rocks of marine origin, were actually volcanic. He attributed their presence to eruptions of ancient volcanoes in an area now far from active volcanism. Arduino also excluded bitumens as volcanic products, showing that they were instead of organic origin.

The geologist who first demonstrated the volcanic origin of basalt was Nicholas Desmarest (1725–1815), opposing Guettard's claims that basalt columns in the Auvergne were of aqueous origin. Nicholas Desmarest, an official of the French Government, served for some time as Inspector-General and Director of the Manufactures of France (Figure 10.5). In 1763 he began studying the region near Mont d'Or in southwest Clermont, where he found prismatic basalt forming hexagonal columns and other regular shapes. Desmarest was familiar with this form of rock from the Giant's Causeway in Ireland, but he could see that here in the Auvergne it was obviously part of a volcanic lava flow.

While climbing the plateau of Prudelle near Mont d'Or, Desmarest noted some loose prismatic columns of dark basaltic rock that had fallen from the cliff above. Tracing the rocks to their source, he found similar columns standing vertically in that cliff, grading upward into the scoriaceous top of the lava flow. His simple observation of the association of columnar basalt as a component of a lava flow stands as a major advance in science.

> I traced out the limits of the lava, and found again everywhere in its thickness the faces and angles of the columns, and on the top their cross-section, quite distinct from each other. I was thus led to believe that prismatic basalt belonged to the class of volcanic products, and that its constant and regular form was the result of its ancient state of fusion. I only thought then of multiplying my observations, with the view of establishing the true nature of the phenomenon, and its conformity with what is found [in the Giant's Causeway] in Antrim—a conformity which would involve other points of resemblance.

It was clear to him that the hexagonal cracks in the lava were the result of cooling and contraction. Desmarest mapped the area from Volvic to Mont d'Or in detail, showing that the lavas had flowed downhill from the volcanic cones, along valleys cut by rivers, and that new

10.5. The French geologist Nicholas Desmarest (1725–1815) was a pioneer in field volcanology. Eighteenth-century engraving (author's collection).

valleys had been subsequently cut along the margins of the lava flow.

Regarding the source of heat for volcanic activity, Desmarest was decidedly a conservative, and followed the earlier ideas of Guettard and de Buffon regarding the combustion of subterranean masses of coal instead of the "fermentation" of pyrites favored by some of his contemporaries. He stated, however,

> I admit that combustible materials are not capable of producing all the effects of volcanoes; but it should not cast any doubt on the necessity of these materials to serve as fuel for volcanic fire. Expanding water in the next instance seems to be the motive force capable of carrying out the acts of eruption, and of producing disturbance accompanying them; but it is no less true that the condition of all the products of subterranean fire shows that coal is the only material suitable not only to achieve their fusion, but also, by its immense quantity and its disposition in the bowels of the earth, to supply the subterranean fires. Moreover, if one examines the vicinity of the craters and funnels by which flame and smoke surge out in frightful clouds, it will be seen that these craters and funnels are covered with immense masses of slag perfectly similar to the residue of the combustion of coal.

Desmarest was a Neptunist at heart and opposed to the idea that primitive crystalline rocks could have been formed by the action of heat. Like Werner, he argued that volcanoes and their products were mere "accidents" that resulted from local melting of primitive basalts of aqueous origin. Because coal was found only in the younger and upper layers of the Earth's crust, it followed that volcanic activity was a relatively recent phenomenon.

Desmarest first presented his revolutionary findings to the French Academy in 1765, to which he was later elected a member. His results were initially published as a brief unsigned article entitled *Basalte d'Auvergne* in 1768, and plates of columnar or prismatic basalt and a report of his discovery that all basalts were of volcanic origin were published in the *Encyclopédie* of Denis Diderot. In 1774 and 1775 Desmarest finally published the details of his research in the *Mémoires de l'Academie Royale des Sciences* in France.

Although Desmarest had unequivocally demonstrated the volcanic origin of basalt, the debate between the Vulcanists and Neptunists continued to rage into the nineteenth century. His theory of basalt as lava was bitterly opposed. One of the most vociferous critics was the Reverend W. Richardson (1803), a Neptunist whose attack involved a peculiar blend of nationalism, religion, and anti-revolutionary rhetoric: "this associate of Voltaire and d'Alembert, in their labours to rid mankind of their religious prejudices, this same M. Desmarest, is now stiled the father of the Volcanic theory, having in the course of the mission on which we have traced him, discovered basalt to be a volcanic production: nor did he neglect to apply this discovery to the main object of his mission, to wit, an impeachment of the credibility of Moses." Desmarest took no part in the controversy, however, and to his opponents he invariably replied: "Go and see."

One contemporary, however, did put Desmarest's hypothesis that basalt was formed by the fusion of granite in the volcanic fire inside the Earth to the test. The Swiss geologist Horace Bénédict de Saussure (1740–99) fused a number of granite rocks and let the melt cool; failing to produce basalt, he concluded that it could not be the product of fusion. Although his deduction was incorrect, de Saussure must be regarded as one of the first experimentalists in geology.

The discovery of columnar basalt as lava was not the only one of Desmarest's contributions to geology. In the late eighteenth century, maps began to be used to communicate geological ideas and information. Perhaps the first of these were the maps that Desmarest drew to illustrate the distribution and occurrence of volcanic rocks in the Auvergne district in 1779. In these maps Desmarest used a subtle range of engraving techniques to show the volcanic cones, lavas, and outcrops of basalts in the area.

The pioneering fieldwork of Desmarest and Guettard stimulated a number of other French and Italian scholars to examine both ancient volcanic formations and active volcanoes more closely. Among them was the French mineralogist and mining engineer Antoine de Genssane, who published a five-volume work on volcanism and heat in the Earth (1776–80).[10] Genssane also proposed that the fire within the Earth was fueled by bitumen, and that this fire—which did not need air—melted rocks and exploded when the molten material came in contact with air at the Earth's surface, causing volcanic eruptions. He discussed the formation of columnar basalt and was one of the first to recognize that this peculiar form was due to the shrinkage of the lava as it cooled.

Jean-Louis Giraud Soulavie (1752–1813), the Abbot at Nîmes, also investigated extinct volcanoes in the districts of Velay, Auvergne, Vivaris, and Provence in central and southern France and strongly advocated the volcanic origin of basalt. He is notable for attempting to establish the chronological succession of volcanic eruptions on the basis of the stratigraphic sequence and preservation of their products.

One of the few scholars who studied active volcanoes in this period was Guy Tancréde de Dolomieu, a Knight of Malta, adventurer, French army officer, and professor in the Ecole des Mines in Paris. The memory of this great geologist is preserved in the calcareous Alpine mountains that bear his name: the Dolomites. As noted previously, during his travels in Italy and Sicily he was the first to recognize volcanic ash deposits interbedded with marine sediments, his most important contribution to volcanology. In 1784 he showed that marine limestones in Sicily contain numerous layers of dark volcanic ashes and basalts, and proposed that submarine eruptions were discharging volcanic products at the time when the limestones were accumulating on the ocean floor. He was the first to demonstrate that basalts sandwiched between sediments need not be derived solely from igneous intrusion, as proposed by Hutton (who referred to them as "unerupted lavas"), but could equally be the products of submarine volcanism.

After his travels in the Lipari Islands in 1783, Dolomieu was convinced that heat alone could not account for the fluidity of magmas. The heat of volcanoes, he maintained, could not come from combustion or the original heat of a cooling star. Although he failed to find another source of heat, he suggested that it was the sulfur in magma that gave it added mobility and exerted an expansive force during eruption, thus providing a bubbly or vesicular structure to lava, scoria, and other volcanic rocks. In 1789 he wrote on the source of heat for the Auvergne volcanics:

> I think that the volcanic agents, as well as the sources of all ejections, lie at great depths within or below the consolidated crust of the earth everywhere, and that the causes which contribute to the combustion accompanying the eruptions and those which produce the fluidity of the lavas remain hidden there. This seems to validate an opinion that I have long sustained, to wit: that the volcanic hearths are not located in the secondary beds as various writers have supposed, that they do not lie in the beds of coal or other combustible materials of vegetable or animal origin, and that if there truly exists a subterranean fire, it is not by that sort of substance that it is fed. In advancing the hypothesis of the fluidity of the center of the globe—or rather in believing in its possibility—and in deducing its likelihood from the phenomena which it would serve to explain, I do not undertake to show or even to indicate the agent, whatever it may be, that prevents the complete solidification of the materials of which it is composed.

With an abundance of active volcanoes in their backyard, Italian scholars continued to have a distinct advantage in the study of volcanic phenomena. Lazzaro Spallanzani (1729–99), a professor of natural history in Pavia, studied several volcanoes and described their form and activity in *Travels in the two Sicilies and some parts of the Apennines*.[11] In some respects, his observations and experiments surpassed all previous studies on the nature of volcanism. Being a good experimentalist, Spallanzani also tested Dolomieu's idea that sulfur played a critical role in the melting and vesicularity or bubbling of volcanic rocks. He showed that, with or without sulfur, the melting behavior of volcanic rocks was identical and that they produced the same type of glassy rock on solidification. Opposing Dolomieu's theory of combustible substances within magmas, he pointed out that flames were never observed on the surface of flowing lava and, furthermore, solidified lavas could be brought back to fluid condition simply by adding heat, without sulfur or other combustible substances. A strong opponent of the "combustion theory," he was one of the first geologists to consider heat as a fluid substance or *caloric* that is transported in magma and lost at the surface during the cooling of lavas.

One of the chapters in Spallanzani's volume bears the title "Experimental enquiries relative to the nature of the gases of volcanos, and the causes of their eruptions." He had studied gases emanating from lavas, and in an experiment collected the gas given off from molten volcanic glasses (obsidian) from the Lipari Islands. This experiment produced bubbles within the molten volcanic rock, caused by an "elastic fluid" (expanding gas) that he suggested might be "the evaporated glass itself," rendered gaseous by the heat. He proposed that this elastic vapor, when trapped in depths of the Earth, produced earthquakes and could break out in volcanic eruptions. He was also among the first to consider the consequences of the interaction of sea water and magma, suggesting in 1798 that sea water might percolate through rocks of the Earth's crust and come in contact with hot materials at depth, resulting in violent effects such as volcanic explosions. On the other hand, he rejected the idea of the ancient Greeks that winds or air moved through passages in the Earth.

Spallanzani had observed columnar lavas in the crater of the island of Vulcano and wrote in 1798:

> The first time I ventured to explore the bottom of the crater of Vulcano, I only found some fragments of this prismatic lava: but when I repeated my visits, and had divested myself of the fear I at first felt, and more carefully examined this dreary bottom, I was enabled to complete my discovery by ascertaining the origin of these prismatic, or, as some may choose to call them, these basaltiform [columnar] lavas. For raising my eyes to that part of the crater which was over my head, and facing the north-east; I perceived a large stratum of lava, almost perpendicular, divided lengthwise into complete prisms, some of which were continued with the lava and made one body with it.

Nevertheless, he was ambiguous on the origin of basalts: "The word *basaltes* is used by Pliny and Strabo to denominate an opaque and solid stone, of the hardness, and nearly of the colour, of iron, commonly configured in prisms, and originally brought from Ethiopia, of which the Egyptians made statues, sarcophagi, mortars, and various utensils. This premised, it remains to enquire whether this stone was of volcanic origin or not, by repairing to the places where it was found, and attentively examining the country to discover whether it bears the characteristics of volcanization. This labour, however, has not, to my knowledge, been hitherto undertaken by any one."[12] He then states that Dolomieu had observed many Egyptian basaltic objects, such as statues and sarcophagi in Rome, and declared that they showed no action of fire: "The green basaltes have never suffered the action of fire," concluding that the Egyptian basalts were "produced by Nature in the humid way" (i.e., by sedimentation). He disagreed, however, with "Werner and other Germans" that all basalts were produced in this manner and concluded that "many are certainly the product of fire" as shown by the prismatic lavas of Vulcano and Felicuda in the Aeolian Islands. "It consequently appears that Nature obtains the same effect by two different ways." Thus, he considered the Swedish prismatic traprocks (columnar basalts) of aqueous origin. He described *shoerl* or scoria at the base of basaltic lavas as added evidence of their volcanic origin and, with Dolomieu, believed that prismatic or columnar jointing in basalts was formed by instantaneous contraction when lavas flowed into the sea and only rarely in lavas flowing on land. Spallanzani also gave an identity to the newly emerging field of science when he referred to the researchers of volcanic activity as the "volcanists."

11

Baron Münchausen in the Volcano

Basalt der schwartze Teufelsmohr,
aus tiefster Hölle
bricht empor

(Basalt, the black devil, breaks out from deepest Hell)

Goethe, 1788

Two notable geologic events occurred in the early part of the eighteenth century, events that were to strongly influence the thinking about the workings of the Earth. In 1707 a submarine eruption in the underwater caldera of Santorini (Thera) led to the emergence of a new volcanic island that continued to grow until 1712. And on November 1, 1755, a major earthquake shook all of western Europe and was felt in parts of Africa, the West Indies, and as far afield as the Great Lakes in North America. In Portugal, the city of Lisbon, closest to the Atlantic Ocean epicenter of the earthquake, experienced the loss of an estimated 30,000 to 60,000 lives and most of its large public buildings were destroyed. Both of these convulsions aroused intellectual curiosity and scientific inquiry about the astonishing power stored in the Earth.

In 1755, the year of the Lisbon earthquake, Rudolph Erich Raspe (1737–94) matriculated at the University of Göttingen in Germany at the age of eighteen. This brilliant figure was to play a major role in the discovery of the volcanic origin of basalt and the birth of volcanology, although he never saw even the steam or smoke of an active volcano.[1] Early on, Raspe displayed a lively interest in geology and was attentive to the discussions of his father, an accountant in the Hannoverian state department of mines and forests, held with the managers of the silver, lead, and copper mines of Clausthal and elsewhere in the Harz Mountains. As a youth he was also introduced to the rocks and the mountains by his father during excursions to the mines of the region. After Göttingen, he took up studies at the University of Leipzig, graduating in 1760. He joined the secretive masonic lodge in Hannover, where he made friends and influential contacts. [2] He was a zealous scholar and published his first geologic work at the age of twenty-six in 1763: *Specimen Historiae Naturalis Globi Terraquei* (A Specimen of the Natural History of the Globe). This work was greatly influenced by the writings of Robert Hooke, who in 1705 had published *Lectures and Discourses of Earthquakes and Subterraneous Eruptions*. That work, which had been largely forgotten, was "discovered" anew by Raspe, who regarded Hooke's ideas as the best explanations put forth on the origin of earthquakes and volcanic activity. In the *Specimen* Raspe advocated the theory of actualism; that is, that the processes that transformed the Earth in the past are the same as those actually observed on the Earth at present. He was particularly interested in the birth and growth of volcanic islands, which he considered a product of uplift, and accounted for the lavas and volcanic rocks as merely a rather superficial covering over the uplifted ocean floor. He mentions columnar basalt in the Giant's Causeway and in Reunion Island in the Indian Ocean "as the rarest kind of stone" having "a certain internal organization and structure," but does not discuss its origin, although he apparently considered it a sedimentary rock at the time. Raspe's publication caused quite a stir and he soon found himself at the center of a large circle of correspondents about geology, including the American Benjamin Franklin, whom he met in Hannover in 1766, and Sir William Hamilton in Naples. With his newfound fame as a philosopher of the Earth, Raspe was promoted to the position of secretary in the Royal Library. His reputation as a scholar increased further when he discovered a neglected pile of unpublished papers of the great German philosopher Leibniz in the Hannover library and published them in 1765.

By 1767 Raspe had been appointed curator of the antique gem, coin, and medal collections of the Margrave of Hesse-Kassel, a post that also included the Chair of Antiquity at the University of Kassel and a seat on the Hessian Privy Council. In Kassel his financial situation was also becoming somewhat easier, and he had more time to tramp around the countryside and study the local geology, particularly in the Habichtswald hills around the Margrave's estate. At this time Raspe heard of Desmarest's great discovery of volcanic rocks in the Auvergne district in France. The revolutionary demonstration that columnar basalts were the products of lava flows from volcanoes had profound effects on Raspe, who did much to try to introduce this Vulcanist view to Germany and Britain.

During his travels in the Hesse region, Raspe came across basaltic rocks locally known as *Wacken* that were commonly used as paving stones in the area, including the streets of Kassel. Tracing them to local quarries, he examined their outcrops with poorly developed prismatic structure and declared them basalts of volcanic origin.[3] This led Raspe to the revolutionary conclusion that the remains of ancient volcanoes existed in the heart of Germany: "We live here on and near the ruins of extinct volcanoes as quietly and securely as we should rest on the most bloody fields of ancient battles, or on the tombs of raging tyrants." His important findings were published in Göttingen in 1771 and the same year in English as *A Short Account of Some Basalt Hills in Hassia* in the *Transactions* of the Royal Society of London. Raspe's paper on German basalts was considered epoch-making by the poet and naturalist Goethe because it first introduced the volcanic origin of basalt to Germany.

The depths of Raspe's early understanding of the volcanic origin of basalts are also well illustrated in his 1769 correspondence with Sir William Hamilton, the patient observer of the eruptions of Vesuvius who had examined more lava flows than any other student of volcanoes at the time.[4] In a letter to Hamilton, Raspe inquired, " I would like to know from your Excellency, if you have ever found among the new lavas of Vesuvius, anything which confirms the interpretation of Mr. Desmarest; in other words, do any of the cooled lava flows of the Vesuvius display toward their end something similar to prismatic basalt?" Hamilton replied, "Concerning the basalt, I have, since investigating lavas, always considered it as a type of lava. However, I have never found any kind which is columnar or polygonal like basalt, among the numerous types of lavas of the Vesuvius, Sicily and the island of Ischia." Raspe also corresponded with other colleagues in Britain and published much both in English and German, with frequent letters to the Royal Society of London. His work was well received in London, where he was elected a Fellow of the Royal Society in 1769.

From his own observations and those of Hamilton, Raspe concluded that while columnar basalt was undoubtedly volcanic and formed when lavas erupted on the sea floor, not as the product of subaerial eruptions, "because its mass cooled faster." In 1769 he wrote: "I was induced to attribute their origin to a watery crystallization which might have taken place, either at the first settling of chaos, or at the time of dissolution of a great part of our globe. I had said the same thing in regard to the Giant's Causeway in my account of the formation of new islands. But I now begin to entertain some doubts about that opinion." His idea that columnar basalt represented submarine eruptions was in part influenced by the half-submerged state of the classic columns of the Giant's Causeway, but here he was wrong, and later work by Faujas de Saint-Fond in the Auvergne (1783) demonstrated the connection between columnar basalt lavas and subaerial lavas. Saint-Fond (1742–1819), a professor at the Museum of Natural History in Paris, had studied older volcanic formations at Vivarais in France and clearly showed the association of columnar basaltic lavas with volcanic cones in his paper *On the Extinct Volcanoes of Vivarais and Velay* (1778). He also visited the columnar basalt formations of the Scottish Isles, demonstrating their volcanic origin in a 1797 publication. And in southern France he found volcanic tuff, identical to the famous pozzolana of the Naples region, and established a flourishing cement industry in France as a result.

Although Raspe's salary as curator of the Margrave's coin and medal collection was relatively generous, bad investments and an extravagant lifestyle left him in severe financial straits. To meet the demands of his creditors and stave off bankruptcy, he resorted to embezzlement. He pilfered from the Margrave's treasures and the coin collection entrusted to his care, and secretly pawned them to Kassel's moneylenders as security for ready cash. In the hope that a generous dowry would solve his financial problems, he married the daughter of a wealthy Berlin doctor in 1771, but continued to sink deeper in debt. This situation caused him to take even more desperate measures. In 1774 Raspe set off from Kassel, ostensibly to Italy on an archeological mission for the Margrave, but never reached his chosen destination. Instead, he made a detour to Berlin to attempt to rescue his impossible financial situation and avert a scandal. However, shortly after his departure his embezzlement was discovered during an inventory at the court, and Raspe was summoned back to face the consequences. After making a full confession, he fled from Kassel to avoid arrest, "with three pistoles in my pocket," went to Clausthal in the Harz Mountains, and then escaped from Germany in April 1775, abandoning his wife and children. Raspe next appeared in Holland in June 1775, and in September of the same year showed up in London. At first he was greeted warmly by colleagues who were impressed by his writings on volcanism and unaware of his crimes. In London he made every effort to join Captain James Cook's third and fateful expedition to the Pacific as the official scientist of the expedition, but fell into disgrace when news of his exploits reached England and the president of the Royal Society. In December 1775 the Society expunged his name from its rolls, the only Fellow "to be expelled on grounds of character."

In January 1776 Raspe was a penniless and landless outcast in London. Undaunted and with characteristic resiliency he promptly applied to the printers of the Royal Society to publish a volume on the progress of German geology. By the summer of that year *An Account of Some German Volcanoes, and their Productions, With a new Hypothesis of the Prismatic Basaltes: Established upon Facts* was published. This work immediately aroused great interest in Britain; Raspe had confronted head-on the heated debate between the Vulcanists and Neptunists on the origin of basalt. On the source of the basalts of the Giant's Causeway he wrote: "The Irish basalts in the Giants-causeway in the county of Antrim appear in polygon articulate columns or prisms. There ought to be some natural reason for that, whatever it be. Is it owing to the difference of their substance and mixture? to the different sloping of the ground on which these fiery melted masses run forward? to the different quickness of their motion? to their different fluidity or is it owing rather to the manner of their refrigeration?" The answer, Raspe proposed, was in submarine eruption: "The regular prismatic columnar black rock or basalt seem, according to their situation and quality in this country, and in many others, to be lavas, which in a hot-fluid state broke their way from underground immediately in the sea, or ran into it, or cooled under its level, without any eruption, within the strata, caverns, and holes, wherein they have been brought to fusion."

At first, Raspe's original work and his English translation of Desmarest's ideas were highly influential, converting many in Britain to the Vulcanist view, but only ten years later his writings were to be largely ignored by the Huttonian school, perhaps as a result of his tarnished reputation. This publication therefore failed to rescue Raspe's academic career and his fortunes continued to decline. He next attempted to make a living in London by translating German works as well as trading in fossils and minerals. By 1782 Raspe had established contact with the English mining engineer Matthew Boulton and his collaborator James Watt, the

11.1. In the late eighteenth century, the German geologist Rudolf Erich Raspe was at the forefront in advancing new ideas about volcanism in northern Europe. His reputation was destroyed, however, by an embezzlement scandal in Germany and he fled to Britain. He died in obscurity in 1794, after having published his most famous but anonymous work: *Baron Münchausen's Narrative of his Marvellous Travels*. The portrait medallion of Erich Raspe at the age of forty-seven is by his Scottish friend James Tassie. Scottish National Portrait Gallery, Edinburgh.

inventor of the steam engine, who employed him as a geologist and mineral prospector in the Cornwall tin and copper mines. Raspe moved to the remote Cornish mining village of Redruth in the winter of 1782. Although a foreigner in those parts, he soon established a solid reputation among the miners as a person very knowledgeable about minerals and mining practice, and his childhood education around the mines in the Harz and Clausthal now stood him in good stead.[5]

While in Cornwall it appears that Raspe continued to write, although not on geologic matters. On a business trip to London in October 1785, he paid a visit to a bookseller in Fleet Street and sold him a manuscript entitled *Baron Münchausen's Narrative of his Marvellous Travels and Campaigns in Russia*. It is reported that he received only a few guineas for the forty-two page book, scarcely enough to pay his coach fare between Cornwall and London.

Thus, Raspe's best known work was in a hugely popular collection of stories about the escapades of a fictional character published anonymously in 1786, but not in the field of volcanology. Apparently he considered that more vulgar works like *Münchausen* were beneath him and reserved his own name for serious works of science. Reviewers generally ignored this new book, except for brief comments on its satirical production, where the marvelous has never been carried to a more whimsical and ludicrous extent. However, the general public soon discovered the Baron as a character synonymous with bravado, which every man has and most men condemn. Raspe had undoubtedly created a character who for popularity and complexity stands beside Don Quixote and Oedipus. Many more editions were soon printed in England, Germany, and France, and ultimately in Boston by 1790. The entire world has since been exposed to the Baron, with wonderful illustrations by such renowned artists as Crowquill, George and Robert Cruikshank, Rowlandson, Gustave Doré, and Théophile Gautier. Although *Münchausen* is a joke, it is one full of bitterness and pain. The Baron is thus a creature of Raspe's own frustration and egotism, and the thundering lies of the Baron recall Raspe's own personal tragedy and downfall (Figure 11.2).

Like Raspe himself, his fictional Münchausen exhibits an interest in volcanoes. In the twentieth chapter he visits Sicily, prompted by an account in the book of Patrick Brydone, A *Tour of Sicily and Malta* published in 1776 (Figure 11.3), and sets out to climb Etna, "determined to explore the internal parts, if I perished in the attempt. After three hours' hard labour I found myself at the top; it was then, and had been for upwards of three weeks, raging." After walking around the edge of the crater, our hero continues: "At last, having made up my mind, in I sprang, feet foremost. I soon found myself in a warm berth, and my body bruised and burnt in various parts by the red-hot cinders which, by their violent ascent, opposed my descent; however, my weight soon brought me to the bottom, where I found myself in the midst of noise and clamour, mixed with most horrid imprecations; after recovering my senses, and feeling a reduction of my pain, I began to look about me. Guess, gentlemen, my astonishment, when I found myself in the company of Vulcan and his cyclops, who had been quarreling for the three weeks before mentioned, about the observation of good order and due subordination, and which had occasioned such alarms for that space of time in the world above." After being restored to health by Vulcan and shown "every indulgence" by his beautiful wife, the goddess Venus (Figure 11.4), Münchausen continues: "Vulcan gave me a very concise account of mount Etna; he said it was nothing more than an accumulation of ashes thrown from his forge; that he was frequently obliged to chastise his people, at whom, in his passion, he made it a practice to throw red-hot coals at home, which they often parried with great dexterity and

11.2. An engraving of Baron Münchausen from the earliest illustrated edition of the book (author's collection).

11.3. The summit of Mount Etna in Sicily, drawn at about the time when Raspe's fictional character Baron von Münchausen explored its internal parts. Engraving 1813 (author's collection).

then threw them into the world, to place them out of his reach, for they never attempted to assault him in return, by throwing them back again....Our quarrels," he added, "last sometimes three or four months, and these appearances of coals or cinders in the world are what I find you mortals call eruptions." Mount Vesuvius, he assured me, was another of his shops, to which he had passage three hundred and fifty leagues under the bed of the sea, where similar quarrels produced similar eruptions....At this stage some tattlers had whispered a tale in Vulcan's ear about Münchausen's relationship with his wife Venus, which roused in him such a fit of jealousy that he dropped Münchausen into a pit so deep that it brought him to the surface on the opposite side of the Earth.

Raspe never acknowledged authorship of the book, but continued his work for Boulton and Watt in the Cornwall mines. However, in late 1786 the mines had hit hard times and Raspe, after losing his position as an assay master, returned to London and disappeared for three years. In the spring of 1789 he appeared in Edinburgh, on his way to prospect for metals and minerals on behalf of the baronet Sir John Sinclair in the Scottish Highlands. Apparently the wandering German prospector left an unfavorable but lasting impression on the Highlanders, and it is generally believed that Sir Walter Scott used Raspe as his model for the character of the villain Hermann Dousterswivel in the novel *The Antiquary*.

Although he was out, Raspe was never completely down. During the next few years he was involved in several unfruitful schemes of prospecting and mining enterprises in Wales, Cornwall, and Scotland, some under the control of his former employer Matthew Boulton. By 1793 Raspe was in Ireland, prospecting for mineral-bearing lands. In November he traveled far to the west, to the derelict copper mine at Muckross in the most primitive parts of Ireland. Here the cold and damp climate along with his failing health made him an easy victim to an epidemic of spotted fever at the age of fifty-seven. He died far from the world of learning and science that he had striven to belong to, but he died in the field, practicing geology, the subject he loved. Raspe is buried in an unmarked grave in the hillside graveyard at Killeaghy Chapel, among the lakes of Killarney. All that remains is a bare entry of his death in the parish register of St. Mary's Church in Killarney.

The Plutonists and Vulcanists

The school of geologists who opposed both the Wernerian view and the Neptunists came to be known as the Plutonists, and their undisputed intellectual leader was the Scottish geologist James Hutton (1726–97). He demonstrated for the first time in 1788 the phenomenon of intrusion of molten rock or magma into layered strata and proposed that this molten rock originated in the highly heated interior of the deep Earth. Hutton's view of the Earth was revolutionary and his conclusions regarding the great age of the Earth, incompatible with Gen-

11.4. When visiting Sicily, Baron Münchausen jumped into the crater of Etna volcano and met the god Vulcan, who ordered his wife Venus to show the Baron "every indulgence." From the 1929 edition of *Travels of Baron Münchausen*, engraved by John Held Jr.

esis, were especially considered a danger to religion. Hutton was, however, an intensely religious man, and he argued that the Earth's processes followed a divine plan, in a peculiar combination of geology and natural theology. Thus, volcanoes existed both as safety valves for the release of excess heat from within the Earth and as a global force because their expansive and uplifting power was needed to raise the surface of the Earth and contribute to the soils so that plants and animals might flourish. Hutton thus saw volcanism as a means toward the realization of a higher purpose in the natural order of things: "A volcano should be considered as a spiracle to the subterranean furnace, in order to prevent the unnecessary elevation of land, and fatal effects of earthquakes; and we may rest assured that they, in general, wisely answer the end of their intention, without being in themselves an end, for which nature has exerted such amazing power and excellent contrivance." Hutton's view of volcanoes, and even his nomenclature, are curiously similar to those set forth earlier in the *Geographia Generalis* (1672) by the Dutchman Bernhard Varenius:

> Burning mountains and volcanoes are only so many spiracles serving for the discharge of the subterranean fire, when it is thus preternaturally assembled. And where there happens to be such a structure on conformation of the interior parts of the Earth, that the fire may pass freely and without impediment from the caverns therein, it assembles unto these spiracles, and then readily and easily gets out from time to time without shaking or disturbing the Earth. But where such communication is wanting, or the passages not sufficiently large and open, so that it cannot come at the said spiracles without first forcing and removing all obstacles, it heaves up and shocks the Earth, till it hath made its way to the mouth of the vulcano; where it rushes forth, sometimes in mighty flames, with great velocity, and a terrible bellowing noise.

Hutton had studied medicine in Edinburgh, Paris, and Leyden, but on his return to Edinburgh did not take up the profession of physician; instead he interested himself with chemical experiments and geological studies. Unlike many of his contemporaries who wrote much about rocks but spent little time examining them, Hutton devoted great energy to the observation of geologic strata in the field, in Scotland, England, and France (Figure 11.5). His field studies led to his paper "Theory of the Earth," which is considered by many a masterpiece in the history of geology. It was presented to the Royal Society of Edinburgh in 1785 and published in the Society's *Transactions* three years later. Although Hutton's work was revolutionary and decades ahead of its time, its acceptance was severely hampered by his very unattractive and obscure style. This attack on the Neptunists also received strong opposition from a number of theologians because they regarded the rejection of the Neptunist dogma as a direct attack on the sanctity of Christian truth.

Hutton and his followers had observed layers in the Earth's crust that, although resembling basalt in mineralogical and other properties, were sandwiched between layers of sedimentary rock. They recognized that these rocks, which they referred to by the common mining term "whinstone," were not volcanic but of igneous origin and had been forcefully intruded between the sedimentary layers. Furthermore, they concluded that these rocks had solidified from a melt or magma. Hutton (1788) recognized that rocks could consolidate from a melt by a process he called "congelation" brought about "immediately by the action of heat." This was the key to his concept of igneous intrusion: "Foreign matter may be introduced into the open structure of strata, in form of steam and exhalation, as well as in the fluid state of fusion [as magma]." He had been convinced that some layers of basalt or whinstone within sedimentary deposits were the products of such igneous intrusion; in 1785 he set out for Glen Tilt in the Grampian Hills of Scotland to investigate this concept. There he found granitic veins cutting through adjacent sedimentary rock strata and concluded that the granite was not "the original or primitive part of the earth." [According to his friend John Playfair (1805), when Hutton made the find, it "filled him with delight, and called forth such marks of joy and exultation that the guides who accompanied him were persuaded that he must have discovered a vein of silver or gold."]

Contrary to the Neptunist theory of precipitation of basalt and other related rocks from a universal ocean, Hutton maintained in 1795 that these siliceous rocks, being neither soluble in water nor porous, could never have been deposited from aqueous solution. Citing the eruption of volcanoes and the fusion of rocks by heat, he deduced that the Earth had a central perpetual heat supply and invoked the action of this heat as a major geological agent.[6] No one before Hutton had been so bold as to suggest a subterranean intrusion of molten rock. Those who adopted this opinion were termed Plutonists and were regarded as carrying out the Volcanist doctrines to the greatest extremes. The Plutonists proposed a deep-seated origin for magma or molten rock and suggested a heat source much further down than the generally accepted conflagration of combustible matter at shallow depths in the crust as proposed by the Volcanists.

11.5. A caricature showing James Hutton (1726–97) in the field, wielding a geologic hammer and confronting a rock outcrop decorated with stony faces. Could these be the visages of Hutton's principal antagonists, the Neptunists? From Paterson, *Kay's Edinburgh Portraits*, 1885 (author's collection).

Yet Hutton's primary concern was with the effects of heat in the Earth as a geological agent, not its causes or origins: "Our knowledge is extremely limited with regard to the effects of heat in bodies, while acting under different conditions, and in various degrees," he wrote in 1788. Unlike his contemporaries, he did not regard heat as a result of fire or combustion of matter within the Earth, but instead as a perpetual heat that in his mind was as fundamental to the Earth as the force of gravitation. Just as the Earth has "a determined quantity of gravitating matter...it also possesses a certain determined quantity of heat." Neither he nor his collaborator John Playfair succeeded in explaining the origin of this heat. Hutton rarely referred to "fire," but frequently spoke of the "power of heat" within the Earth, which was related to his theory of matter. He considered two basic forces in the universe: attraction (gravitation) and repulsion. The latter included specific heat, which influenced volume, (i.e., the expansion of material on melting), and latent heat, which related to fluidity. In Hutton's view, heat was literally immaterial, unlike the material view of his contemporaries who considered heat as matter: *the caloric*. In a sense, his ideas were precursors to the concept of the kinetic theory of heat and the science of thermodynamics.

The Scottish mathematician John Playfair (1748–1819) was Hutton's close friend and strongest spokesman, biographer, and interpreter. In 1802 Playfair published *Illustrations of the Huttonian Theory of the Earth*, which explained, with considerable literary skill and in more lucid prose, the dynamic geology Hutton had proposed. In further developing Hutton's theories, Playfair sought the causes of eternal heat not in combustion but as a primary heat in the nucleus of the globe; he also attributed heat to the effects of the great pressure exerted on the Earth's interior. Playfair was a professor of mathematics and natural philosophy at the University of Edinburgh. He was among the leading figures of the Scottish Enlightenment, a circle that included the engineer James Watt, the chemist Joseph Black, the economist Adam Smith, and Hutton. Together with Joseph Black (the discoverer of carbon dioxide or "fixed air") and Adam Smith, Hutton was a founder of the Oyster Club in Edinburgh, which became an important center of scientific debate. In the eighteenth century, Edinburgh was "the northern Athens," with a literary, intellectual, and academic life that made contemporary Oxford and Cambridge pale in comparison. Joseph Black (1728–99) and Hutton were particularly close associates in this society and were always seen as a pair. Inseparable, they became known as "the abstemious cronies" (Figure 11.6). According to Playfair, "Doctor Black hated nothing so much as error; Doctor Hutton hated nothing so much as ignorance."

In a way, Playfair was to Hutton what Huxley was to Charles Darwin: his own contributions were largely submerged in his advocacy of a larger cause.[7] In his classic work on Hutton's theory, Playfair also tackled the question of the origin of whinstone or trap, found interbedded with sedimentary rocks in many regions or as vertical veins or dikes. He demonstrated the great similarity between whinstone and basalts and stated that "they may safely be regarded by the geologist as belonging to the same *genus*." Whinstone, he concluded, represented "*un-erupted lava*"—that is, magma that has been intruded into the Earth's crust but has not reached the surface. He also documented the columnar internal structure of some whinstone, pointing out its similarity to the columnar structure of some basalt lavas. On the question of volcanic heat, Playfair noted severe difficulties with the theory of subterranean combustion as a source. First, because of the great longevity of individual volcanoes, the required supply of bituminous substances or other combustible matter should be long exhausted. Second, the structure of the Earth's crust does not permit the circulation of air, which is required to fuel such combustion. "If, on the other hand, we ascribe the phenomena of volcanoes to the central heat, the account that may be given of them is simple, and consistent with itself." He regarded the nucleus of the globe as a hot, melted, fluid mass. Fissures, shafts, or tubes issuing upward from the nucleus region carry molten rock, which may solidify as whinstone and basalts in the crust or erupt at the surface as lava. On the origin of heat, Playfair was evasive (1802): "There is no incongruity in supposing the production of heat to be independent of combustible bodies, and of vital air. We are, indeed, in all cases, strangers to the origin of heat."

The rather vague ideas on primordial heat in the Earth espoused by Hutton and Playfair were strongly criticized by their contemporaries, who then ignored them and focused on the irrelevant combustion hypothesis.[8] Most of the attacks centered unfairly on the vast quantities of combustible matter, such as bitumen, coal, or sulfur, that the subterranean fires would demand, as well as the air supply for the combustion, even though neither Hutton nor Play-

11.6. "The abstemious cronies" in the Oyster Club, Edinburgh: James Hutton, considered the father of geology by some (left), and the chemist Joseph Black, the discoverer of carbon dioxide. From Paterson, *Kay's Edinburgh Portraits*, 1885 (author's collection).

fair had at any stage proposed subterranean combustion. The Irish chemist and avowed Neptunist Richard Kirwan was one of Hutton's main critics; in 1793 he strongly protested the idea of a magmatic origin of granite: "Where then will he find those enormous masses of sulphur, coal or bitumen necessary to produce the immense heat necessary for the fusion of those vast mountains of stone now existing?" It is ironic that the term Plutonist was originally suggested by Kirwan, one of the greatest antagonists of the Huttonian school.

Many of Hutton's opponents had maintained that the melting and solidification of crystalline rocks, such as granite or gabbro, would only yield an amorphous, glassy mass on cooling, as observed in glass-making furnaces, and therefore the process could not produce basalt, whinstone, or other volcanic rocks. This objection to the Huttonian theory was then tackled experimentally by the Scottish geologist and chemist Sir James Hall (1761–1832), the youngest of the Oyster Club geologists. Reluctant at first to accept Hutton's theory, Hall decided to test his colleague's hypothesis on the origin of magmas and volcanic rocks. Yet he abandoned his initial experiments on melting and crystallization on basalt in 1790 out of respect for Hutton's open dislike of experimentation. Hutton, a great skeptic of experimental methods in general, maintained that it was wrong to "judge the great operations of the mineral kingdom, from having kindled a fire, and looked into the bottom of a little crucible." He believed that the heat and pressure to which the rocks had been subject in nature during fusion lay far beyond the reach of laboratory simulation and that "the operations of nature were performed on so great a scale, compared to that of our experiments, that no inference could properly be drawn from one to the other." Although close friends, Hall and Hutton did not agree on all scientific matters and took opposing sides in a full-scale debate on the new versus old chemistry in the Royal Society of Edinburgh in 1788.[9] In the debate, the young Hall presented the new theories of Antoine Lavoisier, attacking the phlogiston theory, in which all combustible substances are regarded as containing an inflammable element that is given off during combustion, and demonstrating that combustion and respiration were the result of chemical reactions with oxygen. Hutton, however, defended the old phlogiston theory. It is likely that these diverging views on the fundamentals of chemistry were an obstacle in their collaboration during Hutton's lifetime.

James Hall, descendant of an old Scottish family who succeeded to a baronetcy at the age of fifteen, knew Hutton from his youth, both as a neighbor and family friend. His true love was chemistry, and here he had the instruction of the two founders of the science, first studying under Joseph Black in Edinburgh in 1781–82 and in 1786 under Lavoisier in Paris. Hall also traveled widely in Europe from 1783 to 1786, and on hearing that Vesuvius was in eruption in 1785, his curiosity about volcanism was aroused and he moved south to Naples. During his visit to the famed volcano, Hall was struck by the "vertical lavas" that run up the crater wall of Monte Somma. At first he thought they represented fissures that had been filled from above by the flow of lava, but soon realized that they were dikes or the product of upward flow of magma through fissures in the Earth's crust and thus signified the subterranean feeding channels of volcanism at the surface.

Hall also saw the volcanoes of the Lipari Islands and Sicily in 1785 in the company of the French geologist Dolomieu. The lava samples he collected on these travels were included in his historic melting experiments—whose results were first presented to the Royal Society of London in 1798 and published, under the title *Experiments on Whinstone and Lava*, in the Transactions of the Royal Society of Edinburgh in 1805. To test Hutton's claim that rocks such as whinstone or basalt were derived from originally molten matter, or magma, Hall set out to examine the melting and cooling of basalts.[10]

Hall began his experiments with seven of his lava samples, including some from Iceland, and eight samples of basalt. First he melted the specimens in the reverbatory furnace of an Edinburgh iron foundry and then converted the melted rock to glass by rapid cooling or quenching. Although we do not know the temperatures of the experiments, they were in excess of the melting point of silver at 950°C. As a temperature guide, Hall placed ceramic cones of various sizes and types into the furnace, that he obtained from the famous porcelain manufacturer Josiah Wedgewood. As the cones melted he was able to determine a relative temperature scale in his furnace. In this manner he eliminated the possibility that any crystals inherited from the original rock would remain, biasing the results (this approach is still a standard technique in modern experimental petrology). Next Hall remelted the crystal-free glass and allowed the melts to cool slowly for up to twelve hours at moderate temperatures. In so doing he produced crystalline or partly crystalline rocks, rough and stony in texture, that resembled the original rock samples. These rocks, he showed, had melting and crystallization features exactly the same as those of basalts. Demonstrating that melts of basaltic rocks precipitate silicate crystals on cooling was, of course, of fundamental importance in establishing the volcanic theory on the origin of basalt—and thus a death blow to the Neptunist theory of aqueous origin for basalt, granite, and other rocks of igneous origin. His experiments thus reproduced in miniature the processes that Hutton held responsible for the formation of rocks when he demonstrated that mixtures of melted minerals may be solidified either in glassy, slaggy, or wholly crystalline form by varying the rate of cooling. These findings were the first evidence of the temperature dependence of the texture of igneous rocks.

But if basalt rock was merely a product of solidification from a high-temperature silicate melt, then the experimental fusion and recrystallization of the rock should not affect its chemical composition. To test this, Hall submitted specimens of the products of his melting experiments to Robert Kennedy for chemical analysis in 1805; he found their composition was essentially the same as the original rocks. A further chemical and mineralogic test for a volcanic origin of basalts was carried out by Louis Cordier in 1816, who analyzed lavas from active volcanoes and ancient basalts in detail to see if he could detect any difference in their mineral components. Cordier crushed the rocks, separated the minerals in liquids by virtue of their density, and then analyzed the particles chemically, under the microscope and with magnets. He found no significant difference between recent lavas and the ancient basalts. Hall's demonstration, which was of fundamental importance in establishing the volcanic theory for the origin of basalt, dealt the final death blow to the Neptunist theory of an aqueous origin for basalt, granite, and other rocks of igneous origin.

Hall also made the first high-pressure experiments on rocks to test another of Hutton's theories: that limestone, when heated under pressure, would not decompose, but would recrystallize or melt and crystallize into marble on slow cooling. He made a pressure chamber from an old cannon, inserted some limestone, sealed the ends of the chamber, and by heating the cannon reached a pressure of about 270 bars (equivalent to the pressure at a depth of about 800 m in the Earth's crust) subjected the limestone to temperatures near 1000°C. The experiment demonstrated that carbonate minerals retain their carbon dioxide when heated under great pressure and confirmed Hutton's theory. This demonstration that high pressure prevented the decomposition of limestone was also a critical observation in the debate between the Plutonists and the Neptunists on the origin of basalt. It was well known that limestone decomposed on heating to lime at atmospheric pressures. The Neptunists therefore pointed out that layers of basalt found interbedded with limestone in the Earth could not have been molten because then the limestone would have decomposed to lime, with the emission of carbon dioxide. Hall's experiments helped destroy this argument, by demonstrating that limestone could indeed survive high temperatures so long as the pressure was also elevated. In these experiments Hall was the first to use capsules of the inert metal platinum to prevent reaction between the experimentally produced melts and the iron in the surrounding cannon wall at high temperatures and pressures.

In applying these high-pressure experimental results, Hall pointed out that the ejection of blocks of limestone and gypsum from volcanoes such as Vesuvius was clear evidence that the source of volcanic activity and the region of magma generation was by no means superficial, as proposed by those who regarded volcanism as a consequence of combustion, but rather occurred at considerable depth. This depth had to be sufficient to generate the high pressures necessary for preventing the decomposition of carbonate in the high temperature magmas. On the origin of the heat source, Hall was less certain, but was inclined to either a primordial heat or the heat generated by chemical reactions. Although, like Playfair and Hutton, he generally refers to the heat as "internal fire," Hall was by no means an advocate of combustion and merely used this term for a general heat source:

> The original source of internal fire is involved in great obscurity; and no sufficient reason occurs to me for deciding whether it proceeds by emanation from some vast central reservoir, or is generated by the local operation of some chemical process. Nor is there any necessity for such a decision: all we need to know is that internal fire exists, which no one can doubt, who believes in the eruptions of Mount Vesuvius. To require that a man should account for the generation of internal fire, before he is allowed to employ it in geology, is no less absurd than it would be to prevent him from reasoning about the construction of a telescope, till he could explain the nature of the sun, or account for the generation of light.

Having established the great depth for the origin of magmas, Hall proposed "the igneous formation of all rocks from loose marine deposits." He envisaged the accumulation of great sediment deposits on the ocean floor, leading to their compression and heating from a deep source and eventual melting:

> Beds consisting of a mixture of various substances would be still more affected by the same heat. Such as contained iron, carbonate of lime, and alkali, together with a mixture of various earths, would enter into thin fusion, and, penetrating through every crevice that occurred, would, in some cases, reach what was then the surface of the earth, and constitute lava: in other cases, it would congeal in the internal rents, and constitute porphyry, basalt, greenstone, or any other of the numerous class of substances, which we comprehend under the name *whinstone*.[11]

Although Hall was a pioneer in experimental work on high temperature and high pressure experimental petrology, he was not the first to conduct experiments with molten rocks.[12] The first melting experiments of basalts were done by the Italian scholar Francesco d'Arezzo in 1670, who fused Etna lavas, but it has been proposed that the Frenchman René Antoine F. Réaumur (1681–1757) is the rightful father of experimental petrology.[13] In 1726 Réaumur was the first to study the melting and solidification behaviour of mixtures of earths and minerals in a porcelain oven, but he was driven by a desire to find the magic mixture of the raw materials that could reproduce the coveted Chinese porcelain. During experiments in 1739 Réaumur found that the rate of cooling was a critical factor, and that melts could be returned to a granular condition or devitrified upon slow cooling, thus resembling rock. He also began experiments on the crystallization of metals from melts and came to the important conclusion that nearly all metals contract upon cooling. In 1776 the glass manufacturer James Keir had made the recognition that Réaumur's devitrification of glass was in fact a crystallization proces.[14] James Keir had shown in experiments that the melt of ordinary glass produced crystals if allowed to cool slowly. Following thc important discovcry that crystals could grow from a silicate melt, Keir furthermore proposed "that the great native crystals of *basaltes*, such as those which form the Giant's Causeway or Staffa, had been produced by the crystallization of a vitreous *lava*, rendered fluid by the fire of volcanoes." Joseph Priestley (1733–1804) had also fused Iceland lava in a sandstone retort in 1774, and observed that the melt emitted a gas that consisted in part of carbonic acid. Later Gregory Watt (1804) also showed that upon cooling, heated basalt produced jointing structures similar to those found in prismatic natural basalt. All of these eighteenth-century experimentalists owed a great debt, however, to the industrial glassmaker, who established an art which is one of the oldest and going back to pre-historic times. With the development of a three-chambered furnace and bellows, glassmakers had developed a technology which later enabled these scholars to fuse rocks.

Important contributions to the development of the Plutonic theory were also being made in continental Europe at this time. Most notably, Scipio Breislak (1748–1826) described volcanic rocks and associated phenomena in Italy in his book *Introduzione alla Geologia* (1811). He was a Plutonist, but not of the Huttonian school, and attributed the fluidity of magmas to the joint action of heat and water.

12

Chemical Reactions as the Source of Heat in the Earth

Sir Humphry Davy
Abominated gravy.
He lived in the odium
Of having discovered sodium.

E. C. Bentley, 1905

Sulfur, called *apyron* by the Greeks, was one of the first true chemical elements recognized by our ancestors, along with carbon in the form of charcoal and the metals of gold, silver, and copper. This bright yellow and stony substance, which was readily available around the craters of volcanoes, was found to burn with a strange, sputtering blue flame, giving off evil-smelling and choking vapors like those that belched from volcanoes. It was no wonder that early philosophers attributed volcanic heat to the combustion of sulfur in the Earth, and the idea that this heat was generated by chemical reactions within the Earth had gained strong support by the seventeenth and eighteenth centuries. Even some of the strong advocates of Plutonism in the early nineteenth century, such as James Hall, were uncertain about the relative importance of chemical heat versus primordial heat in the Earth.

Sulfur was used widely for household and medicinal purposes in antiquity. Alchemists were also intrigued by this yellow stone that burns, and they equated sulfur with heat. They prepared sulfuric acid, "oil of vitriol," by heating ferrous sulfate or the mineral pyrite in a retort and collecting the fumes in water. It escaped no one who handled sulfuric acid that it produced large quantities of heat when combined with water.

Isaac Newton was one of the first to propose a connection between the exothermic reactions of sulfur and volcanic heat, as previously discussed. A number of Newton's contemporaries also speculated on the role of chemical reaction in supplying volcanic heat. In 1683, predating Newton's work, L. de Capola proposed that volcanic heat originated in chemical processes. The British physician Martin Lister (1638–1712), famous for his pioneering work on antisepsis and methods of sterilization, called attention in 1693 to the heat given off during the decomposition of iron pyrites, suggesting that volcanic eruptions might be due to this chemical reaction. When sulfur and sand came together, he proposed, the mixture became heated due to fermentation of the sulfur, causing volcanic explosions. For those who maintained that volcanic activity was fueled by the combustion of bitumen, coal, and other flammable material, Lister's idea provided an active method for setting these materials on fire. The French academician and philosopher Nicolas Lemery (1645–1715) gave proof to Lister's hypothesis with experiments on the spontaneous combustion of a mixture of iron filings and sulfur buried in moist soil. Lemery was the author of *Cours de Chimie*, the most popular chemistry text of its time, first published in 1675. In the English translation (1686) experiments with the "fermentation" of sulfur and iron filings are described:

> Take equal quantities of the *filings of Steel*, and *Sulphur* powdered. Mix them together, and make them into a Paste with water; put this Paste into an earthen Pan, and leave it a fermenting four or five hours...the matter does grow so hot of it self, that a man can hardly endure his hand upon it. It is the same thing whether you make a smaller quantity, or make five and twenty or thirty pounds of the Preparation at a time, it flames, and half calcines before it is put upon the Fire, which cannot be explicated but by the violent action and frication of the *acid* part of the Sulphur against the solid body of this metal. This Operation may very well help us to explicate after what manner the Sulphurs do ferment in the earth when it happens to tremble, and fires do burst forth, as does too often happen in many Countries, and among others at Mount *Vesuvius*, and Mount *Aetna*; for these Sulphurs mixing in *Iron* Mines may penetrate the Metal, produce a heat, and at last take flame after the same manner as they do in the present Operation. And it will be in vain to object, that there is no Air in the earth to help to fire the Sulphurs; for there are clefts sufficient in the earth to give entrance unto Air. But if there were not enough, the fermentation which happens at the meeting of *Iron* and *Brimstone* may be able to raise the earth in some places and to burst it a-sunder. The great heat of many Mineral waters may likewise easily be explicated by the means of these Subterranean Fires.

In 1690 Lemery reported on further experiments with sulfur and iron. He buried two parts of wetted sulfur and one of iron filings beneath the ground. Within a few hours they had fermented, become greatly heated, and a flame arose from the surface with slight explosions, bearing a strong resemblance to volcanic eruption. These experiments may represent the first step in experimental geology:

> I take a mixture of equal parts of iron and sulfur powdered; this I form into a paste with water...which swells with considerable heat. The fermentation accompanied with heat, and even fire...seems to me fully sufficient for explaining after what manner fermentations, shocks and conflagrations are excited in the bowels of the Earth, as happens in Vesuvius, Etna and divers other places: for if iron and

sulfur happen to meet together, and are intimately united and penetrate each other, a violent fermentation must ensue, which will produce fire, as in our operation.

Later, the French geologist Dolomieu also proposed that the ignition of sulfur in volcanoes produced the heat that led to fusion of rocks, causing lava flows. Similar views were held by the encyclopaedic George-Louis Leclerc de Buffon. In *Histoire naturelle* (1750) de Buffon related one type of earthquake to the processes occurring at the roots of volcanoes, stating that

> It must be remembered that all substances which are inflammable and capable of explosion do, like powder, at the instant of their inflammation, generate a great quantity of air; that air thus generated by fire is in a state of exceeding great rarefaction, and, from its circumstance of compression within the bowels of the earth, must produce most violent effects. Suppose now that at considerable depth, as a hundred or two hundred fathoms, there should happen to be pyrites and other sulphureous matters, and that, through the fermentation excited by the filtration of waters or by any other means, they come to ignite...These matters taking flame will produce a great quantity of air, whose spring compressed in a small space, as that of a cavern, will not only shake the ground around it, but will attempt all ways of escaping and being at liberty. The passages which offer are the cavities and trenches formed by subterraneous waters and rivulets; the rarefied air will be precipitated with violence into every passage that is open to it and form a furious wind, the noise whereof will be heard on the earth's surface accompanied with shocks and tremors.

The abundance of sulfur in the Earth's crust was the critical element in de Buffon's theory. In regions not troubled by earthquakes and volcanic eruptions, such as England, pyrites must be lacking in the deep crust; in a region such as Italy, prone to earthquakes and eruptions, pyrites must be abundant.

The causes of volcanic eruptions and earthquakes were attributed to both fermentation

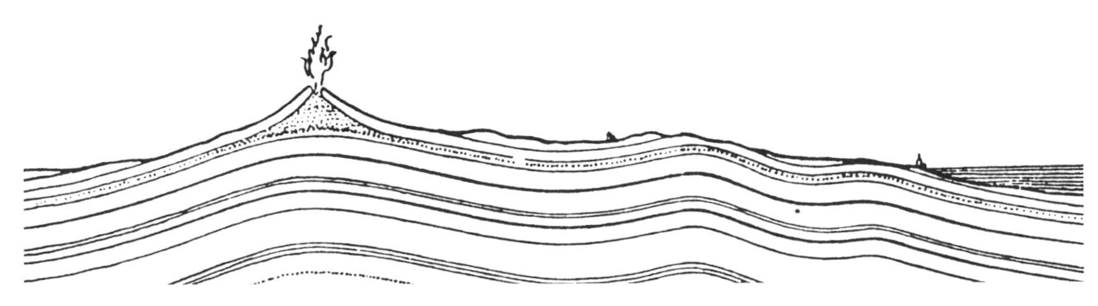

12.1. The Cambridge professor John Michell (1724–93) considered volcanic fires fueled by the combustion of coal beds and "fermentation" of sulfur in the deep Earth. Vapors generated by the heat found their way to the surface and caused eruptions. From Michell's 1761 Royal Society of London paper.

of pyrite and combustion in the Earth by John Michell (1724–93), a professor of geology at the University of Cambridge.[1] His study of the Earth was sparked by the great Lisbon earthquake in 1755. In the publication of his results in 1761, Michell attributed the cause of volcanoes and earthquakes to subterranean fires, taking the conventional view that these fires were fueled by the combustion of coal beds and "fermentation," or reactions in pyritous or sulfur-rich shale within the Earth's crust (Figure 12.1).

Michell also pointed out the proximity of volcanic areas to the sea, proposing that subterranean fires are the cause of underground explosions, which in turn produce earthquakes.[2] "These fires, if a large quantity of water should be let upon them suddenly, may produce a vapour, whose quantity and elastic force may be fully sufficient for that purpose...those places that are in the neighbourhood of burning mountains, are always subject to frequent earthquakes; and the eruptions of those mountains, when violent, are generally attended with them." Thus, Michell considered eruptions or "fires" in volcanic regions to originate in rock strata rich in coal or shale where pyrite is present in sufficient quantity to combust, but that the quantity of pyrite varies in different parts of the strata, so that the fires are not continuous.

Immanuel Kant also became interested in the convulsions of nature after the Lisbon earthquake.[3] In the *Allgemeine Naturgeschichte und Theorie des Himmels*, published anonymously in 1755. Kant proposed, following closely the ideas of Nicholas Lemery put forth fifty years earlier, a number of cavities within the Earth where considerable quantities of sulfur and iron could "ferment" when wetted by water, with the chemical action leading to subterranean conflagrations, upheaval of the Earth's crust, earthquakes, and volcanic eruptions. The subterranean cavities, he maintained, were particularly large and well developed under mountains, with an abundant supply of air to support the conflagrations. Detailing the chemical action, Kant wrote:

> One takes 25 pounds of iron filings, an equal amount of sulfur, and mixes it with common water. Bury this "dough" one or one-and-a-half feet deep in the earth and press it firmly on top. After the passage of some hours, a dense vapor is seen rising; the earth trembles, and flames break forth from the soil. There can be no doubt that the first two materials are frequently found in the interior of the Earth, and water seeping through cracks and crevices can bring them into a state of fermentation. Another experiment produces spontaneously inflammable vapours from the combination of cold materials. Two drams of oil of vitriol [sulfuric acid] combined with eight drams of water, when poured onto two drams of iron filings, bring forth a violent effervescence and vapours which ignite spontaneously. Who can doubt that vitriolic acid and iron particles are contained in sufficient quantity in the interior of the Earth? When, now, water reaches them and brings about their mutual interaction, they give off vapours which try to expand, make the ground tremble, and break out in flames at the orifices of the fire-spewing mountains.

Kant also regarded volcanoes as the "safety valves" of the Earth:

> It has long been observed that a country is relieved from its violent earthquakes if a volcano has broken out in the vicinity, for it is through this means that the enclosed vapours find an exit. And it is

known that around Naples the earthquakes are much more frequent and terrible when Vesuvius has been quiet for a long time. In this manner, that which frightens us often serves in a beneficial way, so that if a volcano were to open up in the mountains of Portugal it could become an early indication that the misfortune was gradually going to disappear.

Kant had proposed in *Allgemeine Naturgeschichte* in 1755 that all matter in the universe was initially distributed throughout space, in a finely divided condition, and that an ordered universe was formed by gravitational forces acting on this vaporous mass. Vortices within led to formation of concentrated masses, and gave rise to the planets. His nebular theory was to have profound influence on our thinking about the interior of the Earth and the retention of primordial heat in the planet. Building on Kant's idea, the French mathematician Pierre Simon de Laplace, proposed a similar nebular theory. In two epoch-making works, the *Méchanique Céleste* on dynamical astronomy and the *Exposition du système du monde* (1796), he presented a new hypothesis on the origin of the planetary solar system from a primordial nebula. He assumed that all bodies of the solar system originated from an immense and incandescent nebula, rotating from west to east. As the nebula cooled, it contracted and rings of matter were formed, all lying in the same equatorial plane. Gradually each of the rings would break up into rotating masses that would eventually coalesce to form the individual planets. These nebular hypotheses strongly supported all ideas regarding an originally hot Earth that was gradually losing heat, and matched well the view among many geologists in the early nineteenth century that volcanic and other igneous forces had been much greater in the distant geologic past than at present, giving rise to a strongly directional view of Earth history.[4]

Another proponent of chemical action as a driving force of volcanism was Franz U. T. Aepinus, who in 1781 proposed that volcanism on Earth was the result of interaction of iron and sulfur.[5] He had some difficulty in applying this concept to volcanism on the Moon, however, because it was generally believed that the Moon was essentially devoid of an atmosphere. How could there be combustion in the Moon to drive volcanic action in the absence of oxygen? At a loss to account for this, he suggested that perhaps the combustible material on the Moon could burn without air.

A new and highly influential hypothesis on exothermic chemical reactions as a driving force of volcanism was developed with the discovery of the alkali metals by Humphry Davy (1778–1829) at the beginning of the nineteenth century. Davy's ideas were to have a long-lasting effect on thinking about heat in the Earth; they also proved very attractive to the Neptunists (Figure 12.2). The new chemical theory did not call for a fluid interior, a vast storehouse of sulfur, or an internal fire in the Earth, and was consistent with the prevailing idea that volcanic activity in the Earth was progressively decreasing. Davy, who made a number of fundamental discoveries in chemistry—and invented the miner's safety lamp—was also a pioneer in thermodynamics. He was among the first to realize that heat could not be considered as matter, but could be regarded as a form of "peculiar motion, probably a vibration of the corpuscles or bodies." Only twenty-three when he began to lecture at the Royal Institution in 1801, he became an instant hit in London and soon was the best-known scientist in Britain.

12.2. The British chemist Humphry Davy was the chief advocate of heat-giving chemical reactions as the source of volcanic activity in the early nineteenth century (author's collection).

Davy already had a great interest in geology, and eventually founded the Geological Society of London with George Greenough in 1807.[6]

In his initial lectures at the Royal Institution, Davy followed the conventional view, dating back to the Middle Ages, that the fuel for volcanic heat was the combustion of coal and the fermentation of the sulfur compound pyrite. He also struggled with several other volcanologic problems, including the origin of basalt and, in particular, its columnar form. He acknowledged that such forms had been observed in the lavas of Etna where they flowed into the sea and was ready to accept the lava flow origin of most basalts. Some basalt dikes, he pointed out, were clearly of igneous origin—such as those at Cockfield in Durham, where the coal in contact with the cross-cutting dike has been highly altered by heat to a form resembling coke. In his ninth lecture to the Royal Institution in 1805, Davy emphasized the importance of pyrite and other sulfur compounds as well as that of coal in accounting for volcanic heat. He cited numerous instances of their thermal effects, such as the tale of the Yorkshire farmer named Wilson "who piled up many cartloads of pyrite in a barn of his own for some secret purpose, perhaps to extract the gold. The roof being faulty and admitting rainwater to fall copiously in amongst them, they first began to smoke and at last to take fire and burn like red hot coals so that the town was considerably disturbed and alarmed."

Although Davy at first may have followed the conventional view regarding the heat source, he soon preferred a chemical explanation. By 1806 he had established that the chemical affinity of the elements is electrical in nature, and in 1807, using this principle, was able to isolate by electrolysis for the first time the metals of the alkaline earths—potassium and sodium and later calcium, strontium, magnesium, and barium—and to demonstrate the elementary nature of chlorine. He isolated potassium and sodium by passing a current from a powerful battery through potash and soda in a moist condition. The high heat liberated with the oxidation or "burning" of these metals on reacting with water, with spectacular flames and explosions, so impressed Davy that it led him to propose a new hypothesis for volcanism in 1808. Rejecting the idea of a permanent central heat in the Earth as an explanation, he wrote "If the interior of the globe had been from all time in a state of ignition, the effects must have been long ago communicated to the surface, which would have exhibited not a few widely scattered volcanoes, but one glowing and burning mass."[7]

After his famous chemical discoveries in 1807 to 1811, Davy was convinced that the heat given off during the oxidation of the alkalis and alkaline earths was the source of heat for volcanic action, supposing the existence of large quantities of the metallic alkaline earths within the Earth under volcanic regions:

> The metals of the earths cannot exist at the surface of the globe; but it is very possible that they may form a part of the interior, and such an assumption would offer a theory for the phenomena of volcanoes, the formation of lavas, and the excitement and affects of subterranean heat, and would probably lead to a general hypothesis in geology. Let it be assumed that the metals of the earths and alkalies, in alloy with common metals, exist in large quantities beneath the surface, then their accidental exposure to the action of air and water, must produce the effect of subterranean fire, and a produce of earthy and stony matter analogous to lavas...The interior of the globe is composed of metals of the earths, which the agency of air and water might cause to burn into rocks, and even the reproduction of these metals may be conceived to depend upon electrical polarities of the earth.

Davy illustrated his lectures on the theory of the chemical origin of volcanic heat with spectacular displays, as reported by John Paris: " I remember with delight the beautiful illustration of his theory exhibited in an artificial volcano constructed in the theatre of the Royal Institution. A mountain had been modeled in clay, and a quantity of the metallic bases introduced into its interior; on water being poured upon it, the metals were soon thrown into violent action—successive explosions followed—red hot lava was seen flowing down its sides, from a crater in miniature—mimic lightnings played around, and in the instant of dramatic illusion, the tumultuous applause and continued cheering of the audience might almost have been regarded as the shouts of alarmed fugitives of Herculaneum and Pompeii."[8] Davy would rivet the attention of his audience by placing a small piece of potassium on the surface of a basin of water, thus producing a fire.

In 1813 Davy planned to visit the volcanoes of the Auvergne in central France to test his theories, and obtained a passport from the Emperor Napoleon I for himself and his assistant Michael Faraday. But they never reached the Auvergne because on arriving in Paris, Davy heard of the discovery of a new substance that had been extracted from seaweed. Not to be outdone by his French competitors, he immediately set up a laboratory in his hotel room and soon recognized that the chemical in question was a new element: iodine. The rest of their stay in France, therefore, was devoted to the study of seaweeds and extraction of the new element. He did, however, reach Naples, and saw Mount Vesuvius and, after studying the region, concluded that there must be communication between the Italian volcanoes. He proposed that the caverns underneath Solfatara volcano were linked with Vesuvius, and that when Vesuvius was active, Solfatara was in repose. When he threw a slip of paper into Solfatara's crater, and it was not rejected, he surmised that there must be a current of air descending into the crater to feed the oxidizing and heat-producing chemical reactions beneath Vesuvius.

Still, Davy continued to be of two minds about the origin of basalt.[9] After his consideration of "all the most remarkable volcanoes in the south of France" he wrote in 1814, "All the basalt that I have seen between the Alps and Pyrenees is decidedly of igneous origin. I have observed some facts on this subject that are, I believe, new—a regular transition of lava into basalt, depending upon the different periods of refrigeration; and true prismatic basalt in the interior of an ancient lava." After a trip to Ireland, Davy is again in doubt. In 1823 he wrote, "I have been visiting some of the wildest spots in Mayo and Donegal, and have again been studying the mysterious basaltic arrangements of Antrim [where the Giant's Causeway is located]; but I almost despair of any adequate theory to account for the phenomena."

Davy also made many later observations of Vesuvius, in the spring of 1815, in December 1819, and in January and February 1820, that resulted in his paper *On the Phenomena of Volcanoes*, read before the Royal Society in 1828. By this time he had devoted a significant part of a fifteen-year period in testing his chemical theory of volcanic heat. Yet by 1828 he had to abandon his hypothesis because all the experimental evidence for chemical combustion in Vesuvius was negative. He found, for example, that niter thrown on the hot lava stream produced no visible change in its appearance and, further, when some molten lava was placed in a closed bottle and allowed to cool, the oxygen consumption was not measurable, contrary to his theory. Although still convinced of the important role of the reactions of the alkalis and alkaline earths, in his later years Davy was faced with the mounting evidence for high temperatures within the deep Earth. His final words on the subject reflect this alternative opinion on the origin of volcanic heat:

> On the hypothesis of a chemical cause for volcanic fires, and reasoning from known facts, there appears to me no other adequate source than the oxidation of the metals which form the bases of the earths and alkalies; but it must not be denied that considerations derived from thermometrical experiments on the temperature of mines and on sources of hot water render it probable that the interior of the globe possesses a very high temperature; and the hypothesis of the nucleus of the globe being

> composed of fluid matter, offer a still more simple solution of the phenomena of volcanic fires than that which has just been developed.

In the third dialogue of his *Consolations in Travel* Davy completely surrendered his hypothesis, concluding that volcanoes were due to a central fire in the Earth.[10] Other scholars had, however, proposed similar thermal sources, most notably the French physicist André Marie Ampère (1775–1836), who observed that the exothermic reaction between potassium and water could proceed in an explosive manner, resembling a volcanic eruption.[11]

One of the fiercest critics of Davy's theory about volcanic heat was Gustav Bischof (1792–1870), a professor of chemistry at the University of Bonn in Germany. The arrangement of volcanoes in great continent-wide lines on the Earth's surface, as first pointed out by Humboldt in 1822, was taken by Bischof as a clear indication of the deep-seated origin of volcanic action, ruling out any superficial source within the crust.[12] In his view, the insufficiency of hypotheses regarding a chemical origin of volcanic heat by combustion was self-evident, but he believed Davy's discovery of exothermic reactions during the oxidation of the alkali metals in the Earth deserved another look, perhaps with modification. Davy had maintained that air could circulate through the Earth via volcanic craters and thus participate in heat-giving oxidation reactions. But no less an authority than the French chemist L. J. Gay-Lussac (1778–1850) had shown by sheer logic that it was impossible for atmospheric air to enter deep into volcanoes, because the pressure of the high-density magmatic liquid is acting outward.[13] He witnessed many explosions of Vesuvius during his stay in Naples in 1805 (in the company of Alexander von Humboldt and Leopold von Buch) and realized that the magmatic pressure within the volcano was sufficient to maintain a column of magma equivalent to the height of Vesuvius, about 1000 m (320 ft) above sea level equivalent to the pressure of 300 atmospheres. Air could not possibly flow into the volcanic system and fuel the oxidizing reactions against such a steep pressure gardient.

If air is absent from the volcanic system, could water then be the oxidizing agent of the alkali metals? Bischof noted that steam is normally emitted in great abundance from volcanoes so that water must be present in the volcanic system. However, the reaction of water with the "earthy and alkaline metals" in volcanoes should result in the evolution of enormous quantities of hydrogen gas—a phenomenon that had never been observed. On these and several other criteria, Bischof rejected the chemical theory of heat in the Earth and concluded: "We may, therefore, look upon the hypothesis which seeks the cause of volcanic phenomena in intense chemical action as untenable."

Bischof next considered the hypothesis that volcanism is due to the interior of the Earth being at "white heat," inherited from the time of its formation.[14] He argued that if the temperature increase observed with depth in mines and bore holes is projected downward, then at some depth the rocks in the Earth must be in a state of fusion. "But since they possess such various degrees of fusibility, the more fusible rocks must be in a liquid state, at depths in which the less fusible ones are still solid." He proposed that "the seat of volcanic action" in the Earth is a region where masses of more easily melted rocks are enclosed in solid and more refractory rock, and that the degree of fusibility is directly related to the chemical composition of the rocks, especially their alkali content. The pioneer British seismologist Robert Mallet was also critical of Davy's theory, pointing out that rocks of the Earth's crust contain no more than 4 to 5% of the alkaline metals, which is insufficient to account for the high heat. He wrote: "Had it not been for the splendor of Davy's genius and the announcement of this view at the moment of his great discovery of the bases of the alkalies, it would probably never have had even a momentary acceptance."[15]

After rejecting Davy's hypothesis, Mallet advanced a new theory of heat generation based on mechanical forces and frictional heat in the crust of a contracting Earth. He was also opposed to the idea of a largely liquid nucleus or core, and a thin, solid crust, and pointed out that such a view of the Earth had been largely dismissed by the work of the British geophysicist William Hopkins, who showed that the thickness of the solid crust is in excess of 1280 km (800 mi). Thus, liquid rock could exist only in relatively isolated pockets or "lakes" within the Earth and not constitute a large global molten reservoir. Mallet considered it significant that as the Earth is losing heat (cooling), it must also be a *contracting* globe. Contraction and decrease in volume of the Earth would lead to shrinking of the outer layer or the crust, throwing up great folds and deep depressions, much as the wrinkles on the surface of a dried prune. Differential contraction of the hotter core and rigid crust would lead to great tangential compressive stresses in the crust, which generated heat where the crust fractured. The production of heat would be greatest along major fractures in the crust, where it could lead to fusion of rocks and generation of magmas. Thus, Mallet believed that the real origin of volcanic heat in the Earth does not come from primordial heat, but rather from mechanical work during compression in the cooling Earth's crust. From his experiments on the crushing of rocks, he claimed that the heat produced during crushing was equivalent to about 99% of the work expanded, (i.e., that almost the entire mechanical work reappears as heat).

A strong defender of Davy's views was Charles Bridle Daubeny (1795–1867), who carried out studies in chemistry and volcanology at Oxford University. He had studied volcanic formations in central France in 1819 and mapped this region before the important study of George Poulett Scrope, discussed below. In his book *A Description of Active and Extinct Volcanos* (1826) Daubeny also attempted to explain volcanic heat chemically, as the result of the exothermic reaction of intruding surface water with metallic calcium and sodium, which he thought might be present in great quantitites at depth. In an article in the *Encyclopaedia Metropolitana* (1835) Daubeny endeavored to show that silica and alumina, when combined with either hydrogen or with alkalis, would lead to combustion, and that this chemical action was manifested in volcanoes. He thus attributed volcanic activity to chemical reaction related to the access of water and atmospheric air to bodies in the earth that were capable of absorbing oxygen. Gustav Bischof strongly criticized Daubney's views, leading to an intense debate on the chemical theory of heat in the Earth.[16]

Another of Davy's supporters was the influential geologist Henry Thomas De la Beche

(1796–1855), who was to become the first Director of the Geological Survey of Britain. Although De la Beche favored the idea of a hot interior of the Earth and a "former igneous fluidity" of the planet, he appealed to the oxidation of sodium and potassium in a metallic nucleus of the Earth to generate the heat, and his "volcanic theory" relied on the concept of the Earth's having a largely metallic nucleus, with abundant alkalis and alkaline earths near the surface. Water and other oxidizing fluids migrating toward the nucleus would result in exothermic reactions with sodium and potassium, giving off localized heat that was manifest at the surface as volcanic eruptions. In *Researches in Theoretical Geology* (1834) De la Beche proposed that "a large proportion of phenomena observable in volcanic eruptions appear best to accord with the theory of the percolation of water, containing certain substances in solution to the metallic bases of the earths and alkalies."

Charles Lyell, in his *Principles of Geology* (1830) wrote the most influential text on Earth science in the nineteenth century, but it was a work that did nothing to advance the study of the origin of heat in the Earth. He accepted the Plutonist view of the origin of granite as an igneous rock, formed by crystallization from molten rock at depth. Lyell envisioned a vast store of heat within the Earth, maintaining that at some distance beneath the surface, the Earth was molten under great pressure. Clearly influenced by Davy's electrolysis experiments, he envisaged a cyclical process of chemical action in the Earth that gave rise to powerful electrical currents that produced volcanic heat; this heat in turn gave rise to chemical reactions. He pointed out, in particular, the possible role of oxidation of the alkalis and alkaline earths as a chemical heat source, closely following Davy's ideas:

> Instead of an original central heat, we may, perhaps, refer the heat of the interior to chemical changes constantly going on in the earth's crust; for the general effect of chemical combination is the evolution of heat and electricity, which, in their turn, become sources of new chemical changes. It has been suggested, that the metals of the earths and alkalis may exist in an unoxidized state in the subterranean regions, and that the occasional contact of water with these metals must produce intense heat. The hydrogen, evolved during the process of saturation, may, on coming afterwards in contact with the heated metallic oxides, reduce them again to metals; and this circle of action may be one of the principal means by which internal heat, and the stability of the volcanic energy, are preserved.

Lyell also suggested, tentatively, the fermentation theory as a possible fundamental cause of volcanism and heat in the Earth.[17]

Lyell's theory was decidedly cyclical and very much in the tradition of the Uniformitarian school, implying an equal intensity in the rates of volcanism on the Earth through geologic time.[18] Such a view was strongly objected to by the British geologists John Phillips and Daubeny in 1845:

> Mr. Lyell...supposes that the calorific energy of the interior of the Earth is constantly acting, so as to reconvert sedimentary into crystalline aggregates, *equal quantities* in equal times, and thus to maintain a perpetual equilibrium between the liquifying internal and the solidifying external agencies of the

> Globe. This speculation is much too poetical to be examined according to the dry rules of Baconian Philosophy: if the heat expanded in this operation can be obtained from chemical processes, these must gradually tend towards equilibrium; if from a general internal reservoir of caloric, that reservoir must become less prompt in supplying the incessant demand: either of these effects operating through *indefinite time* must cause the gradual refrigeration of the surface of the Globe.

Lyell's views and Daubeny's criticism demonstrate fairly well how opinions on heat in the Earth diverged during the first half of the nineteenth century. On the one hand there were those geologists who embraced the nebular hypothesis of Laplace and considered current volcanism as the vestiges of the residual heat of the Earth's fiery beginnings.[19] They believed that the geologic record showed evidence of a diminishing igneous force with time, and that the Earth was running out of heat. Opposing this directional interpretation of Earth history was the Uniformitarian view of the Huttonian philosophy, with Lyell its strongest defender.[20] In the absence of evidence for an energy supply other than residual heat, Lyell was at a distinct disadvantage in the debate, which may explain why he devoted only three out of 479 pages in his 1830 book to the discussion of heat in the Earth. Grasping for an explanation, Lyell proposed such alternate energy sources as combustion in the crust and chemical reactions as possible explanations for the continuing heat supply. One skeptic of this idea of chemical heat as a source of volcanism was the British philosopher of science and professor of mineralogy at Cambridge, William Whewell: "The supposition that this fire may be produced by intense chemical action between combining elements, requires further, not only some agency to bring together such elements, but some reason why they should be originally separate." [21]

Davy's hypothesis was so well known in Europe that it became a major feature in the plot of the first science-fiction novel, Jules Verne's 1864 classic *Journey to the Center of the Earth.* The descent of Professor Liedenbrock's party through the crater of the Snœfellsjökull volcano in Iceland revealed that only the outer part of the Earth's crust was at high temperature whereas the deep interior was relatively cool. This led the professor to conclude that the heat in the Earth was superficial, caused by reactions between the alkali metals and water in the outer crust, and not due to deep-seated heat. This enabled the Professor to penetrate to great depths, without encountering increasing temperatures, until his raft was carried upward through a vertical shaft at terrifying speed, first on a wave of water and finally on a fountain of hot magma, until they were shot out of the crater of Stromboli volcano. Verne was a close friend of the French geophysicist Charles Sainte-Claire Deville, who had supplied him with up-to-date information about the hypothesis of Davy and other matters regarding the deep Earth interior.

In 1875 the German geologist A. Mohr argued from evidence of the geothermal gradient that heat in the Earth was due to chemical processes.[22] Mohr's conclusions were based on temperature observations made in a 1023-m (3390-ft) deep boring at Speremberg in Germany (Figure 12.3). Mohr maintained that if the Earth's interior was still molten, then the thermal gradient should become steeper with depth (i.e., the rate of temperature increase should accelerate). We now know that this assumption is wrong; the rate of cooling or heat loss for any hot object such as the Earth increases as the surface of the planet is approached. Mohr noted the rate of temperature increase fell off with depth and extrapolated that at a depth of only 1575 m (5170 ft) the temperature increase would become nil. He regarded this finding as strong evidence against the Plutonist theory of heat, concluding that

> The cause of the increasing heat in the interior of the earth must lie in the upper strata of the earth's crust.... The theory of volcanoes must of course adapt itself to the above results, and the fluidity of the lavas is not a part of the incandescence (no longer) present in the earth, but a local evolution of heat by sinkings which have always been produced by the sea and its action upon solid rocks, as indeed all volcanoes are situated in or near the sea. This local superheating of the volcanic foci contributes greatly to the internal heat of the earth. For the internal nucleus of the earth can lose but little heat outwards on account of the bad conductivity of the siliceous and calcareous rocks, whilst, in the lapse of ages, it must propagate uniformly all the heat-effects of the volcanoes, and thus a constant elevated temperature must prevail in the interior, and therefore we come to the conclusion that increase of heat in the interior of the earth which is everywhere met with is the result of all preceding heat-actions, uniformly diffused by conduction in the internal nucleus of the earth.

Mohr also considered heat produced in the Earth by "sun-warmed, infiltrated fluids, and also chemical processes such as the evolution of carbonic acid by the contact of oxide of iron with

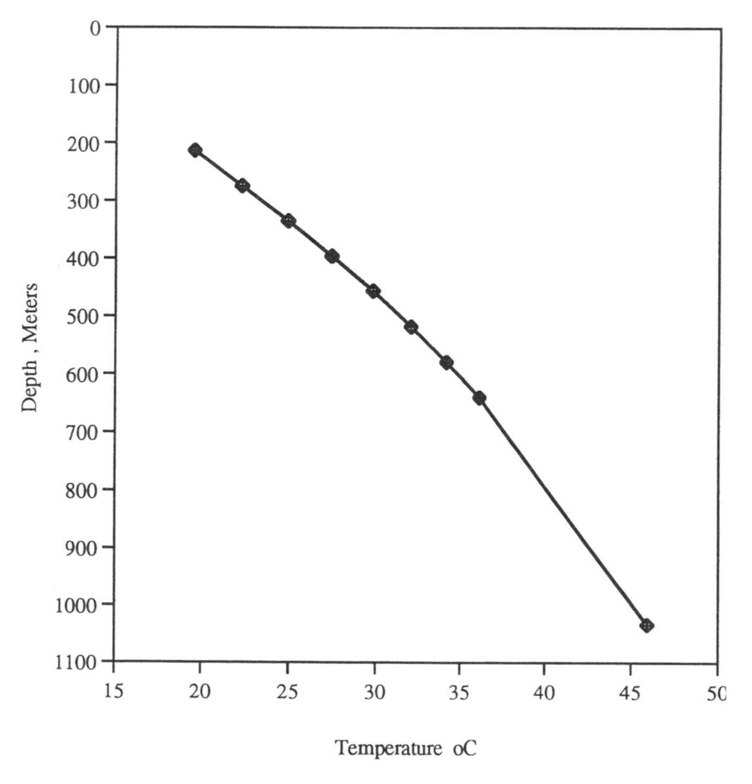

12.3. One of the first to report on the geothermal gradient in the Earth was the German geologist A. Mohr in 1875, based on his measurement of temperature in a 1023-m deep bore hole in Speremberg in Germany. The graph shows the steady increase in temperature with depth.

the remains of organisms, the formation of pyrites and blendes by the reduction of sulphates in contact with organic matters, the decomposition of lignite and coal."

Even near the end of the nineteenth century the chemical theory was still alive. One of its proponents was John W. Judd, a professor of geology at the Royal School of Mines in England, who traveled widely in volcanic regions.[23] He considered the Earth largely solid, possibly to its center, and rejected other scenarios, such as the idea of a thin crust and a largely liquid interior. Judd also rejected the idea of a solid core, solid crust, and a liquid intermediate region, as a result of cooling and solidification from both the center and the exterior, leaving pockets of melt, as proposed by the British physicist William Hopkins. Instead, he favored a solid Earth, from surface to center, with volcanic activity resulting from local fusion of rocks to produce magma due either to pressure changes or chemical and mechanical effects related to frictional heating. Judd was also in favor of chemical action to produce heat, and discussed Davy's hypothesis of a chemical heat origin for melting. In this regard, he did not consider the lack of observation of hydrogen in volcanic gases or the problem of introducing oxidants to the alkali metals at depth as a serious objection to Davy's hypothesis.

Among those who disagreed with Davy was Lord Kelvin (earlier William Thomson, 1824–1907), who was strongly opposed to suggestions that the Earth's heat was regenerated by internal chemical reactions. He considered such proposals violations of the principles of natural history:

> It would be very wonderful, but not an absolutely incredible result, that volcanic action has never been more violent on the whole than during the last two or three centuries; but it is as certain that there is no less volcanic energy in the whole earth than there was a thousand years ago, as it is that there is less gunpowder in a "*Monitor*" after she has been seen to discharge shot and shell, whether at an equable rate or not, for five hours without receiving fresh supplies, than there was at the beginning of the action. Yet this truth has been ignored or denied by many of the leading geologists of the present day, because they believe that the facts within their province do not demonstrate greater violence in ancient changes of the earth's surface, or do demonstrate a nearly equable action in all periods.[24]

The chemical theory was attacked by T. G. Bonney (1899) on different grounds.[25] In disagreeing with Daubeny, who considered the Earth as consisting of unoxidized bases with only the outermost layers oxidized so far, Bonney argued that water penetrating downward would result in oxidation, and that the heat produced in the reaction would be sufficient to melt the adjacent rocks. Such a process would be gradual and incapable of producing a large rise in temperature. Similarly, the predicted flames of hydrogen resulting from the decomposition of water are not generally observed in the craters of volcanoes and, finally, the vast mass of water required is unlikely to be present in the roots of volcanoes.

The American chemist and geologist Thomas Sterry Hunt (1828–92) proposed a radically new concept of volcanism in 1867, based on his ideas of metamorphism of sedimentary rocks.[26] When deeply buried sediments were heated in the presence of water, he suggested, gases were liberated, giving rise to volcanic exhalations. "All volcanic and plutonic phenomena," he wrote, "have their seat in the deeply buried and softened zone of sedimentary deposits of the earth and not in its primitive nucleus. We have seen how actions dependent on water and acid solutions have operated in the primitive plutonic mass, and how the resulting aqueous sediments, when deeply buried, come again within the domain of fire, to be transformed into crystalline and so-called plutonic or volcanic rocks."

Yet despite these many objections, the chemical theory of volcanism was surprisingly long-lived, and continued to be entertained by prominent scientists well into the middle of the twentieth century. Arthur L. Day, Director of the prestigious Geophysical Laboratory of the Carnegie Institution in Washington, D.C., was the last proponent of the chemical theory when he proposed in 1925 that the chemical reaction between gases plays a leading part in generating volcanic heat.[27] In his view, various gases from different sources in the Earth meet within the crust and their reactions lead to local fusion and generation of magma. The Geophysical Laboratory was to become the most important center for research and experimentation on the origin of magmas, largely due to the efforts of Day's colleague, the American geologist Norman L. Bowen. Day had observed the eruptions of Kilauea in Hawaii in 1912 when a 200-m (656 ft) wide and bubbling lava lake occupied the volcano's summit crater. This incandescent pool of molten lava was kept in a state of agitation by the release of gases from the magma, which produced numerous gas fountains at the surface, numbering up to 1100 on one occasion. Day was particularly impressed by the large amount of gas released and its role in influencing the physical properties of the magma and even, he thought, supplying its heat.

Day's idea of heating of the magma as a consequence of the gas release was quite radical. He claimed that "an increase in the amount of gas discharged at the surface of the lake was always accompanied by a rise in surface temperature....This observation immediately led us to the conclusion that there is a causal relation between the amount of gas discharged through the lava and the temperature of the fluid lava itself. The gases somehow contribute to the heat of the lava body." Day pointed out that the gases collected in the lava lake and analyzed by the early American volcanologist E. S. Shepherd contained a variety of species or gas types, which were not in equilibrium as they escaped from the hot lava and "that the gases are necessarily in process of reaction at the time of their release. From our knowledge of gas reactions it follows, further, that gases still in process of reaction at the time of release after passage up through the basin of liquid lava, must have been in process of reaction throughout their upward process, that is to say, there must have been bubbles of gas of different composition uniting beneath the surface at frequent intervals during this excursion." He pointed out that free hydrogen, sulfur dioxide, carbon dioxide, and free sulfur were all present, indicating the highly reactive nature of the gas emissions: "Our knowledge of the reactions which must be taking place between these gases is adequate to establish the fact that heat is being contributed to the lava mass in consequence of these reactions going on within it. We have therefore certainly hit upon one of the sources of energy which serves to maintain the particular type of volcanism which is familiar to all at Kilauea." Day then argued that the heating of the magma is largely a surface feature due to the release and reaction of gases and that the temperature of magma diminishes with depth. "In any event, such temperature distribution does not afford support for the old hypothesis that volcanoes are 'safety valves' to insure the stability of a molten interior, indeed they afford not the slightest indication of the existence of a generally molten interior." Day considered a volcano "a purely local phenomenon arising from unusual local conditions," and that "gases reacting among themselves, contribute materially to the heat necessary to maintain a small lava basin like that at Kilauea in a liquid state."

This idea was particularly appealing to Sir Harold Jeffreys, a leading English geophysicist in the early twentieth century. He had rejected the leading theory of melting in the Earth due to decompression, but maintained that the Earth was a rigid body, incapable of flow or convective motion because of its high viscosity. Because of his belief in a static Earth, he was to become a lifelong opponent of the theory of continental drift and one of the strongest spokesmen against the theory of sea-floor spreading. In his classic text *The Earth, its Origin, History and Physical Constitution* (1952) Jeffreys devoted only two pages to volcanism.[28] He admitted that temperatures in the crust and upper mantle are everywhere lower than the melting points of common magmas and "the generation of liquid granite and basalt is therefore a definite problem." In attempting to minimize the problem he stated, "I do not think, however, that the difficulty is so serious. In the first place, we cannot admit that fusion in the upper layers is a normal condition. Vulcanism is local and occasional, not perpetual and world-wide." In accounting for the "local and occasional" eruptions, Jeffreys appealed to Day's chemical theory of heat generation and melting in the Earth, but completely ignored the geological evidence of abundant volcanism throughout the Earth's geologic history to maintain his view of a static Earth.

13

From Fluid to Solid Earth

Every volcanic eruption, in fact, is the means by which the caloric emanating from the interior of the globe passes off into outer space through the surrounding atmosphere, each volcanic vent acting as a safety valve to the globe.

George Julius Poulett Scrope, 1825

Most geologists studying volcanic activity in the early and mid-nineteenth century continued to envision a planet with a solid crust and a largely molten interior. Heat within the Earth was considered a residue of the central, primitive heat held by the Earth at the time of its formation, a supply that had been diminishing over geologic time.[1] Many considered heat a fluid or gaseous substance, supposedly present in the pores between atoms in matter and caused by thermal expansion. This invisible substance was referred to as *caloric*, and this view persisted with some geologists well into the nineteenth century, when the caloric theory was replaced by the science of thermodynamics.[2] One of the first proponents of the idea that heat is a fluid was William Cleghorn in Edinburgh (1779), but Lavoisier was the first to refer to this mysterious substance as the *calorique* (1787). Spallanzani (1794) also considered caloric as a quintessential substance emitted by the Italian volcanoes,[3] as did Cordier in France.

The pioneer English volcanologist George Julius Poulett Scrope (1797–1876) considered the formation of magma the result of the passage of caloric by conduction from depth to upper levels in the Earth, where the caloric led to melting of rocks (Figure 13.1). Scrope was one of the leading field volcanologists of the

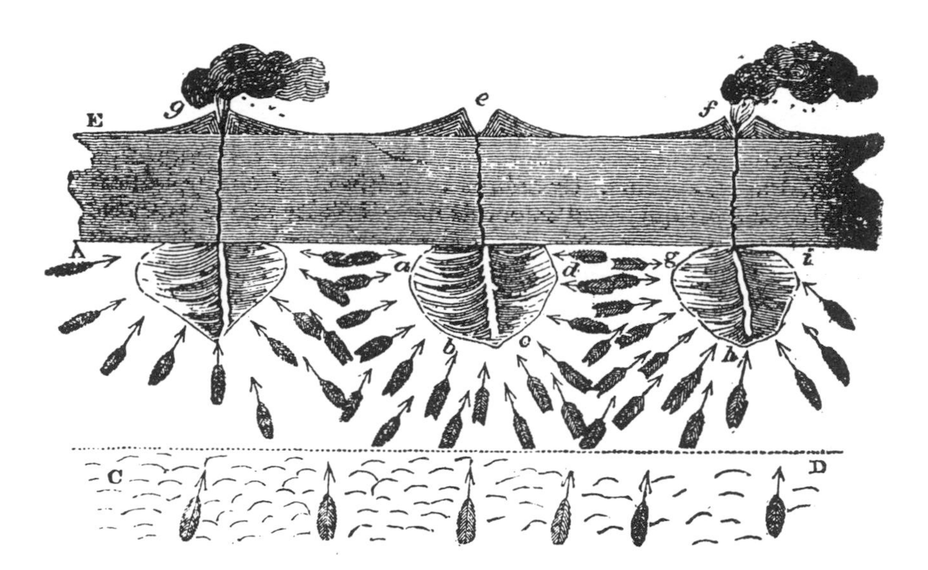

13.1. In 1825 the volcanologist George Julius Poulett Scrope proposed that a heat-giving material, called caloric (feathered arrows), streamed from the Earth's interior up to the roots of volcanoes, generating magma that supplied eruptions at the surface.

early nineteenth century (Figure 13.2). In 1825 he published an important work, *Considerations on volcanos*, and much of the factual material on volcanism presented in Lyell's 1830 classic *Principles of Geology* stems from this source. Here Scrope was quite confident in his view of the immediate causes of volcanic activity: "There can be little doubt that the main agent...consists in the expansive force of elastic fluids struggling to effect their escape from the interior of a subterranean mass of *lava*, or earths in a state of liquification at an intense heat." These "elastic fluids" Scrope considered mainly steam and some other minor volcanic gases. With regard to the heat content of the magma, Scrope believed it to possess that mysterious and invisible substance called caloric, which flowed out of lavas on cooling at the surface. He viewed the escape of steam as the principal means of heat loss or of the flow of caloric out of the lava into the atmosphere, leading to solidification. From the increase of temperature with depth in mines, Scrope concluded that at great depth the Earth was at an intense heat, and that this large accumulation of caloric led to its continued flow toward the surface so that the heat could attempt to attain equilibrium.[4]

As early as the middle of the seventeenth century the learned Jesuit priest Athanasius Kircher had remarked on the increase in temperature of the air with depth in Hungarian mines; mining officers in Schemnitz had told him that they had nothing to suffer either from heat or cold in the mines as long as there was good circulation, but when circulation was poor, the air in the mine invariably became warmer.[5] On the basis of Kircher's and similar observations, the French scientist Jean Mairan (1678–1771) proposed the existence of a fire in the interior of the Earth (1732). But the first systematic measurements were probably not made until 1740 when Antoine de Gensanne showed that the temperature in the mines of

13.2. A portrait of George Julius Poulett Scrope. Woodcut from *Geological Magazine*, 1870 (author's collection).

Girmagny and Vosges in France increased at a rate of about 1°C for every 36 m (108 ft) of depth. Near the end of the eighteenth century, Horace Benedict de Saussure, who had made similar observations in the salt mines of Bex in Switzerland, concluded that there was a steady progression in temperature in the interior of the Earth, at a rate of about 1°C for every 75 m (225 ft) of depth. Many similar studies were conducted in the early nineteenth century in France, Mexico, South America, Cornwall, Saxony, and the Urals, up to 450 m (1350 ft) in depth, with similar results. Bore holes for artesian wells had reached depths of up to 790 m (2370 ft) in Europe by the mid-nineteenth century and showed a similar thermal gradient, on the order of 1°C for every 66 m (198 ft) of depth.[6] Measurements in wells and caves in Paris also demonstrated an increase in depth; this potential heat source was so impressive that it was suggested that the Jardin des Plantes could be heated with water derived by boring to great depths under Paris. When nineteenth-century naturalists extrapolated this gradient to the deeper Earth, they concluded that the planet was occupied by a liquid incandescent mass, an internal ocean of fire, or *Phlegeton*, underneath the crust, at a depth of no more than twelve leagues below the surface. Tides in this ocean were held responsible for earthquakes by the French savant Alexis Perrey (1807–82), in his *Histoire des tremblements de terre* (History of Earthquakes); he proposed that when the tides were strong enough, the liquid fire would break through the terrestrial crust, causing volcanic eruptions.

The transport of heat within the Earth by conduction and convection can give much information about the temperature distribution in the planet's interior as well as the source of the heat. The French physicist Baron Jean Baptiste Joseph Fourier (1768–1830) first developed the theory of heat conduction and applied it to the geothermal gradient in the Earth. In 1807 Fourier, who had gained fame as one of Napoleon's generals in Egypt, recognized the tendency of heat to flow toward an equilibrium; he also established, in his *Théorie analytique de la chaleur* (1822), the principles of the dissipation of heat from a solid body into vacuum or into the atmosphere, in the case of the Earth.[7] Fourier had a strong interest in the geophysical applications of the theory of heat conduction and stated that the geophysical problem had been a prime source of motivation in his research.[8] He discussed terrestrial temperatures and heat flow within a sphere and developed an elegant mathematical analysis of the theory of conduction and the transfer of heat based on the assumption that the flux of heat was proportional to the gradient of temperature. Solar rays, he showed, could only heat the uppermost 100 m (300 ft) of the Earth's crust and, if there were no internal source of heat, the temperature would be constant with depth to the center of the Earth. Yet those measurements taken in caves, mines, and bore holes indicated another source of heat at depth, either from the time of Earth's original formation or some other cause. The fundamental aspect of heat transfer is that heat flows in the direction of decreasing temperature until equilibrium is reached. Fourier's law describes this basic relationship for conductive heat transport, and states that the heat flux q (the flow of heat per unit area per unit time) is directly proportional to the temperature gradient:

$$q = -K \frac{dT}{dy}$$

where K is the coefficient of thermal conductivity and y is the coordinate in the direction of the temperature variation. Fourier developed another formula of heat transfer that can yield the time required for a body such as the Earth to cool to its present temperature:

$$t = \frac{b^2}{\pi \Delta^2} \frac{CD}{K}$$

where t is the time required for a hot sphere to cool down from an initial temperature b to the present temperature (zero), with the observed temperature gradient $\Delta = dT/dr$, where r is the radius of the Earth and T the temperature. The ratio CD/K is the heat capacity and thermal conductivity of the sphere. On the basis of this equation, Fourier extrapolated backward, arriving at a figure of about 200 million years as the time required for the Earth to reach its present temperature. But in those days such a long time seemed highly improbable among geologists and he did not publicize his results or pursue this line of reasoning further. Fourier concluded later that "the Earth preserves in its interior a part of that primitive heat which it had at the time of the first formation of the planets," and that this heat is gradually dissipated through the surface.[9] The temperature gradient he opted for was about one degree for every 30 to 40 m (100 to 120 ft) of depth in the Earth's crust, and the heat flow through the Earth's surface per century was sufficient to "melt a column of ice, of which the base should be a square meter, and height three meters."

Pierre Cordier, who had analyzed lavas for their mineral content (Chapter 11), also collected measurements on the temperature gradient in mines and showed that the temperature in the Earth's crust rose by about 1°C for every 30 to 40 m (100 to 120 ft) of depth. However, he was one of the first to measure the temperature of the rocks, rather than air temperature in the mine shafts, to avoid the influence of air currents. Given this rate of increase, all common rocks would eventually melt at depth, and Cordier concluded in 1827 that the Earth was molten below a depth of 50 km.

Cordier believed the deep Earth contained an abundance of caloric, and that the loss of this substance had gradually led to the consolidation of the globe.[10] He attributed volcanoes to the action of a high-temperature fluid heated by caloric remaining in the Earth's interior. The contraction of the Earth, as a result of its gradual cooling, was sufficient to press molten rock through any openings in the crust, thus producing eruptions. He calculated that a contraction of the globe, equivalent to shortening of the Earth's radius by only one millimeter, would squeeze out enough magma to supply 500 of the greatest known volcanoes. After studying volcanic deposits in Teneriffe in the Canaries and the Auvergne, Cordier estimated that the typical volcanic eruption emitted no more than a cubic kilometer of magma. Since this corresponds to less than a hundredth of a millimeter of the radius of the globe, it would require very little contraction to squeeze such a volume out of the Earth's interior. At the time the prevailing belief was that volcanism had been much more vigorous in the distant past and was continually decreasing—a belief that was, of course, in keeping with the gradual refrigeration of the Earth and fitted well with Cordier's scheme of volcanic phenomena.

In 1835 the French scientist S. D. Poisson pointed out that the crustal geothermal gradients observed by Fourier and Cordier were so steep, that, if extrapolated to the center of the Earth, temperatures on the order of a million degrees would be indicated.[11] At such temperatures the deep interior would be gaseous, which is out of the question, as Poisson noted. His explanation for the heat observed at the Earth's surface was cosmical—that is, that the Earth had at some time in its journey through the universe passed through a very hot region of space, where it heated up superficially. Although he did not doubt that the Earth was originally fluid, judging from its spherical form, he maintained that long ago the planet had lost its primordial heat (1837). During its cooling, however, the outer and cooler portions had sunk into the interior, where they were solidified by high pressure. This was one of the earliest statements on the effects of pressure on the melting or solidification of rocks, a topic of crucial importance in understanding the origin of magmas that we will return to in Chapter 14.

The possibility of a gaseous nucleus of the Earth was an idea that had been entertained early on by Benjamin Franklin. In a letter to the Abbé Giraud Soulavie in 1782, he speculated about heat in the Earth and its internal constitution, pointing out that great uplift of parts of the Earth's crust "seemed...unlikely to happen if the earth were solid to the centre. " He "therefore imagined that the internal part might be a fluid more dense, and of greater specific gravity than any of the fluids we are acquainted with."[12] Franklin's conclusion was that the Earth possessed a solid outer shell, surrounding a hot, fluid interior. The density of air increases greatly with pressure, he noted, and "the dense fluid occupying the internal parts of the globe might be air compressed. And as the force of expansion in dense air when heated is in proportion to its density; this central air might afford another agent to move the surface, as well as be of use in keeping alive the subterraneous fires." Franklin was also a keen observer of rocks, and knew, for example, about the volcanic origin of columnar basalt near New Haven. Franklin's major and lasting contribution to geology, however, was his discovery in 1784 of the atmospheric effects of volcanic eruptions.[13] While serving as an ambassador in France, he observed that an aerosol or dust veil from the 1783 eruption of Laki in Iceland was reducing the solar radiation reaching the ground and proposed that volcanic eruptions could in this manner bring about cooling and influence climate.

Another strong proponent of the idea of a molten Earth interior was Amos Eaton (1776–1842). Eaton studied law and natural philosophy and entered the real estate business in New York, where he was found guilty of forgery in 1811 and sentenced to hard labor for life. The young lawyer had been framed by clients, and because of his exemplary literary and scientific pursuits while in prison, he was granted full pardon after four years. He then enrolled in Yale University to study geology and chemistry under the famous American geologist and mineralogist Benjamin Silliman. It has been said that Amos Eaton was Silliman's greatest discovery. While earning a living as an itinerant geological surveyor and chemical lecturer, Eaton formed a friendship with the powerful and rich Stephen van Rensselaer, whom he persuaded to found the Rensselaer Polytechnic Institution in Troy, New York, to train young men as chemists and mining engineers, men who then might help to exploit America's rich mineral deposits. Eaton's ideas about geology were influenced by the writings of Abraham Werner, but he did recognize basalt as a product of volcanic activity.[14] He proposed great subterranean reservoirs of combustible substances that burst into flames, elevating the mountain belts and fueling volcanoes. This view of exploding matter and expanding gases was a widely accepted theory of mountain building in the United States.

With greater exploration of the Earth and more extensive recording of global natural phenomena, nineteenth-century scholars were faced with rapidly increasing data on the occurrence of earthquakes and volcanic eruptions. Just as their forerunners as far back as Hellenistic times, these naturalists were convinced that these phenomena were intimately related and many claimed that a synchronism existed between violent eruptions and great earthquakes. Charles Darwin, during his epic five-year voyage on the *H.M.S. Beagle*, became convinced of an interplay between volcanoes and earthquakes. After experiencing firsthand the great 1835 earthquake in Chile, he claimed that several volcanoes in South America began to erupt or intensified their activity following that event, and in 1840 compiled records that he regarded as evidence of the interconnection of these phenomena in the Earth: "When first considering these phenomena, which prove that an actual movement in the subterranean volcanic matter occurred almost at the same instant of time at very distant places, the idea of water splashing up through holes in the ice of a frozen pool, when a person stamps on the surface, came irresistibly before my mind. The inference from it was obvious, namely, that the land in Chile floated on a lake of molten stone." These earthquakes, he proposed, "are caused by the injection of liquefied rock between masses of strata."[15] Darwin's idea may have been influenced by the English scholar John Herschel, who had proposed in 1837 "the pyrometric expansion of rocks as a motive power" in geologic processes, such as uplift of landmasses, folding, and faulting. He believed the temperature in the Earth increased with depth "in a frightfully rapid progression," but he steered clear of further speculations about the origin of the primordial heat.[16]

Robert Mallet (1810–81) had observed in 1848 that the areas adjacent to volcanic regions are most prone to earthquakes and he considered the two intimately related. In his 1850 paper, Mallet proposed the "intrusion of subterranean lava" as the mechanism for generating an earthquake. This would cause a rapid condensation of steam and an explosion, with the steam rushing out through fractures produced by the magmatic intrusion, in turn producing an earthquake. Mallet believed the Earth's crust was no less than 45-mi thick, resting on a "liquid melted matter." He supported his volcanic theory of earthquake origin with a compilation of many instances when earthquakes have accompanied or preceded volcanic eruptions.[17] His greatest and generally unrecognized accomplishment was to define in map form a global distribution of earthquakes and active volcanoes following the structural lineaments we now recognize as the boundaries of the major tectonic plates on the Earth.

The same idea was also fostered by Giuseppe Belli in Italy (1850) and by the American geologist Henry Darwin Rogers (1842 and 1856), who carried out fundamental geologic studies of the Appalachian Mountain chain, deducing, like Darwin, a great lake of molten rock beneath the Earth's crust. He proposed that gases expanding below magma-filled cavities in the subcrustal regions of the Earth threw up folds in the overlying crustal rocks. And in France Elie de Beaumont, in 1828 and 1852, who was aware of experiments showing that molten rock shrank on cooling, also proposed that the Earth was originally a molten body with a now-solidified outer crust. Mountain building, he believed, was ultimately due to the shrinking of the Earth as it cooled. The influential American geologist James Dwight Dana (1813–95) adopted de Beaumont's view that the Earth was originally molten. After serving as a midshipman on the *U.S.S. Delaware* in the Mediterranean, where he visited Vesuvius, he joined the Wilkes Expedition to the Pacific, taking the opportunity to study Pacific volcanoes and their products. Dana's view was that the Earth's interior is still liquid and that, in cooling, the Earth's crust had contracted, throwing up mountain ranges as wrinkles between the continental and oceanic areas. Volcanoes were formed at the margins of continental and oceanic areas, in regions of greatest disturbance, where fractures were generated through lateral pressure: "Volcanoes," he wrote, "are indexes of danger," never "safety valves."[18]

In later years Dana modified his views to take into account the evidence of a largely solid Earth presented by the English geophysicist William Hopkins.[19] Dana also accepted the reasoning that the high pressures within the Earth are likely to lead to solidification: "The possibility of solidification at center from pressure, in the face of temperature too high for consolidation from cooling, has not been experimentally demonstrated. Yet a number of facts favor this principle. It has been urged that since the solidification of rocks is attended by contraction, that is, by increase of density, and since pressure tends to produce this greater density, therefore pressure may bring about the condition of the solid." Dana's world now consisted of a thinner molten layer, sandwiched between a rigid central mass that had solidified because of high pressure, and a crust solidified by cooling. This did not, however, require a modification of his views of mountain building as a consequence of the Earth's contraction. Volcanic activity broke the surface where great longitudinal fractures rifted the crust, and the location of volcanoes near the sea had nothing to do with the "vicinity of salt water, but because these were the regions of greatest disturbance and fractures through lateral pressure." All the volcanism in the region we now refer to as the Pacific ring of fire he attributed to "an undercrust fire-sea," Dana's concept of a shrinking planet and a fluid interior was also central to the influential grand synthesis of the German geologist Eduard Suess on the evolution of the Earth's crust, which was popular well into the early part of the twentieth century.

Discovery of a Solid Earth

Knowledge of the deep interior of the Earth, the vast and inaccessible region where magmas evolve and terrestrial heat is stored, became possible only when geophysical measurements at the Earth's surface could be made. The first geophysical studies of the Earth's interior related to its magnetic field, and they in turn had an interesting bearing on ideas regarding heat in the Earth: they provided evidence of a fluid core. It was an early navigational instrument—the magnetic compass—that led, eventually, to experiments with magnetism in the Earth. The compass had been in use for several hundred years when in 1576 the English compass-maker Robert Norman, after noticing the inclination of magnetic needles from the horizontal, experimented by floating them in water.[20] He found that they had a tendency to dip at the north end. The declination, or the deviation of the magnetic needle from true north, was also determined in London at the same time. These observations were to become the basis of a grand theory formulated and published as *De Magnete* in 1600 by the physicist William Gilbert (Queen Elizabeth's personal physician). Fascinated by the lodestone (the natural mineral we call magnetite), Gilbert showed that a spherical piece of lodestone, which he called little Earth or *terrella*, had a magnetic field rather similar to that observed for the whole Earth.[21] He then proposed that the Earth was like a lodestone, with an all-pervasive magnetic field, concluding that the Earth was a giant magnet because it contained a good deal of iron in its deep interior. Gilbert suspected that the natural attraction of objects to the Earth (gravity) was in some way connected to magnetism, and that even the atmosphere was held in place by this force. After learning of the heliocentric picture of the solar system put forward by Copernicus, he even went so far as to suggest that the planets must be held in their orbits by some kind of magnetic power.

Initial studies showed that the magnetic declination appeared to change with time, but this was at first attributed to a poor experimental method. However, in 1634 Henry Gellibrand demonstrated without doubt that the declination of the Earth's magnetic field had decreased by almost 2° since 1622 and that it exhibited a slow westward drift. This astonishing observation showed that the source of magnetism was a dynamic force and led the English astronomer Edmund Halley to suggest in 1691 that the density and magnetism of the

Earth indicated an iron core, and that the drift suggested motions in a fluid or molten interior. A very high density of the Earth's interior had also been inferred from astronomical observations. By 1687 Isaac Newton had estimated that the bulk density of the Earth was about five to six times that of water, based on the gravitational attraction between planetary bodies. Thus, by the end of the seventeenth century it had been established that the deep Earth was exceptionally dense and that its core was turbulent—or at any rate mobile.[22]

Other information about the Earth's interior came from studies of gravitational attraction at various places around the planet. Newton had already predicted in 1687 that mountains, because of their large mass, would deflect a plumb line (a line perpendicular to the Earth's horizontal surface). This prediction led to systematic gravity surveys in the mid-nineteenth century that brought new revelations about the structure of the Earth's interior and its heat. In 1847 the Great Indian Survey published measurements across the Himalaya Mountain's between the cities of Kalianpur and Kaliana, 600 km apart. The head of the survey, Sir George Everest, noted a discrepancy between astronomical and triangulation measurements, which amounted to 5.236 seconds of arc, or about 153 m (460 ft). This error came to the attention of the Archdeacon of Calcutta, John Henry Pratt (1809–71), a mathematician who argued that it was due to the gravitational attraction of the plumb bob used to level the astronomical instruments, an attraction caused by the very large mass of the Himalayas. But in 1856, when Pratt estimated the total mass of the mountain range according to Newton's theory of gravitation, he arrived at a difference about three times greater than that observed. Thus, the great mass of the Himalayas, sitting on top of the crust, did not exert as much gravitational force as might be expected.

The problem was then tackled by the Astronomer Royal of Britain, George Biddell Airy (1801–92), who showed that the discrepancy could be accounted for by assuming that the mountain range had a crustal root of material of similar density that projected into the denser, hot and liquid "lava" interior of the Earth. He imagined the Earth as a spheroid with a crust over an interior of "fluid of greater density than the crust." Airy's "roots of mountains hypothesis" was put forward in 1855: "I conceive that there can be no other support than that arising from the downward projection of the Earth's light crust into the dense lava. It appears to me that the state of the Earth's crust lying upon the lava may be compared with perfect correctness to the state of a raft of timber floating upon water; in which, if we remark one log whose upper surface floats much higher than the upper surface of others, we are again certain that its lower surface lies deeper in the water than the lower surface of the others." Airy's Earth, then, was similar to the cooling star that Descartes and Leibniz had postulated in the seventeenth century.

If the interior of the globe were molten, it should yield to the gravitational attraction of the Sun and the Moon and there would be little or no relative movement of the Earth's surface and the ocean's—that is, no sea tide. The response of the Earth to tidal forces became a critical test for this hypothesis. The French physicist André-Marie Ampère was the first to point out that observations of tidal action did not support the theory. In his *Théorie de la terre*

13.3. The British geophysicist William Hopkins, who pioneered the concept of decompression melting as the primary process of formation of magmas in the Earth. From a portrait in the Department of Mineralogy and Petrology, Cambridge University.

(1833) he wrote: "Those who assume the liquidity of the interior nucleus of the Earth seem to have forgotten the action which the moon would exert on this enormous liquid mass, an action which would result in tides similar to those of our oceans but much worse, as much by their extent as by the density of the liquid. It is hard to imagine how the envelope of the earth could resist the incessant beating by a kind of hydraulic lever 1400 leagues in length." Ampère also proposed that the Earth had cooled in layers around a solid core, and that chemical reactions between the layers caused volcanic heat.

Two prominent Cambridge scholars, William Hopkins and Lord Kelvin, devoted much of their careers to studying the effects of the gravitational attraction on the Earth's shape. Hopkins (1793–1866), a mathematician, was one of the first scientists to carry out research in geophysics and is justly regarded among the founders of that field (Figure 13.3). He was certainly the first to approach geologic problems in a quantitative manner, and it may well be that Archibald Geikie had the work of Hopkins and his students in mind when he wrote in 1868: "Hitherto want of accuracy and definiteness have often been brought as a charge against geology, and sometimes only with too much justice. We seem now to be entering, however, upon a new era, when there will be infused into geological methods and speculation, some of the precision of the exact sciences." As a private tutor and teacher of scientists such as Pratt, Stokes, James Maxwell, and Lord Kelvin, Hopkins's influence on the development of science and in particular on geophysics was great.[23] His interest in geology was fostered by his friend Professor Adam Sedgwick in about 1832 and he sought to elevate it to a higher rank among the physical sciences by applying mathematical and geometrical laws. For this more precise and rigorous science he proposed the term *physical geology*.

From 1839 to 1842 Hopkins analyzed the effects of the Moon and the Sun on the rotational axis of the Earth. He studied the Earth's nutation (nodding or wobble of the axis of rotation of a planet toward and away from the ecliptic pole) and the precession (migration of the axis of rotation) of the equinoxes to determine whether the Earth moves as a rigid sphere or as a solid shell with a liquid interior that can flow with respect to the outer shell. If the solid crust were quite thin, as was generally thought at the time, then the effect of the gravitational attraction of the Sun and the Moon would differ considerably from that observed in the Earth's nutation and precession of equinoxes. In his first paper on the topic (1839), Hopkins showed from studies of the Earth's rotation axis that the solid crust extends to at least a quarter of the depth of the planet's interior; he concluded, from progressively more sophisticated mathematical models published in 1840 and in 1842, that the outer rigid crust must be at least 1500 km in thickness: "Upon the whole, therefore, we may venture to assert that the minimum thickness of the crust of the globe which can be deemed consistent with the observed amount of precession, cannot be less than one-fourth or one-fifth of the earth's radius."[24]

These findings led Hopkins to draw these conclusions regarding the source of volcanic heat in the Earth:

> Many speculations respecting volcanos have rested on the hypothesis of a direct communication, by means of the volcanic vent, between the surface and the fluid nucleus beneath, assuming the fluidity to commence at a depth little, if at all, greater than that at which the temperature may be fairly presumed to be such as would suffice, under merely the atmospheric pressure, to fuse the matter of the earth's crust. When it is provided, however, that that crust must be several hundred miles in thickness, the hypothesis of this direct communication is placed, as I conceive, much too far beyond the bounds of all rational probability to be for an instant admitted as the basis of theoretical speculation. We are necessarily led, therefore, to the conclusion that the fluid matter of actual volcanos exists in subterranean reservoirs of limited extent, forming subterranean *lakes*, and not a subterranean ocean.

By 1839 Hopkins had realized that increasing pressure can cause a hot liquid to crystallize or solidify; consequently the very high pressures within the Earth might solidify even the hot core material. This crucial observation meant that even though the Earth's interior might be considered very hot, it could be largely solid. Hopkins also maintained that the expansive force of primitive heat within the Earth was the principal cause of elevation of mountains and accounted for the action of volcanoes and earthquakes.[25] His major contribution was the demonstration that the Earth models proposing a thin crust and a largely liquid interior were wrong.

The work Hopkins began was taken up by the Reverend Osmond Fisher (1817–1914), who also argued that the great pressure in the interior dictated that it is largely solid.[26] He pointed out, however, that extrapolation of the observed near-surface geothermal gradients required a "fluid substratum" of molten rock beneath the crust. As Fisher's studies progressed, his estimates of the thickness of this fluid layer decreased. He proposed that cooling had resulted in contraction of the planet and calculated that the reduction in volume of rock solidifying from magma was about 6%. Volume contraction of the Earth led to wrinkling of the shrinking outer crust and the formation of mountain ranges. In 1872 he proposed that magmas were generated by "fusion through a diminution of pressure due to the partial support of the mountains by the lateral thrust which has upraised them." This hypothesis assumed a solid crust of tremendous strength, sufficient to support its own weight over large distances. Future work, however, was to show that this crust is weak and, in his 1881 model, Fisher proposed that the Earth had solidified both from outside, forming the crust, and from the center, forming a solid nucleus.[27] The solidification of the exterior was retarded because the action of convection currents continued to bring heat to the surface. Only when convection ceased did the crust begin to form. The solid nucleus was caused, on the other hand, by the high pressure. These processes led to a largely rigid planet, with a thin, molten layer that supplied magma for volcanic activity, sandwiched beneath the solid 40 km (25 mi) thick crust. Fisher considered the layer of magma largely superheated and water-rich, "kept in a state of compression by the superincumbent pressure of the crust, but ready to burst forth with the evolution of steam and gas wherever and whenever a vent is opened for its escape."

Further studies on the gravitational attraction of the Sun and the Moon were carried out by Hopkins's former student, Lord Kelvin. Kelvin's rise in science was meteoric. He published his first paper in mathematics at the age of sixteen, and by age twenty-two had been named a professor of natural philosophy at Cambridge. Kelvin had studied the deformation of a homogeneous, incompressible, and elastic sphere, such as the Earth, when subject to body forces or gravitational attraction. In 1862 he first showed that the response of the Earth to the Moon and the Sun is what would be expected if the planet consisted throughout of a material more rigid than glass—in fact as rigid as solid steel. So the Earth must behave as a solid and highly rigid body on the time scale of its daily rotation around the Sun. By 1863 his studies led him to conclude that the solid crust was no less than 2500 miles in thickness, twice as thick as Hopkins had proposed. [28] Kelvin's discovery of a rigid Earth was a great surprise to geologists. No longer did they have a great molten nucleus beneath a thin crust as a ready source of magma.

Kelvin was fond of demonstrating the effect of fluidity on rotation by spinning a hard-boiled egg and a raw egg in front of his audience during lectures. Although the hard-boiled egg could twirl standing on its axis, the raw egg flopped around horizontally. It was known that the Earth had only a slight nutation, like the nodding of a top on its axis. From this he deduced that, like a hard-boiled egg, the Earth had a high degree of rigidity. And the idea of an Earth with a liquid interior was untenable based on the evidence of sea tides. He held that the lunar tides, which cause the oceans to rise and recede daily, would equally act on and displace the molten interior, resulting in no relative change in land and sea level during the gravitational attraction of the Moon and the Sun. Because the Earth's surface did not respond to the diurnal pull of the moon, its interior must be solid. In its infancy, however, the Earth was a molten globe in Kelvin's view, and its interior rose and fell with the lunar tide, cracking the young and brittle crust and causing the sinking of great crustal blocks into the molten interior where they foundered because of their greater density. This process would have the effect of carrying the cooler crust continually down toward the Earth's center, allowing the hot molten material to flow up toward the surface. The result, over tens of millions of years, would be gradual cooling and solidification of the Earth from the center out, until it reached its current and completely solid state. Later studies showed, however, that on longer time scales (more than 10,000 years) even the Earth's interior is capable of flow.

One of those rejecting Hopkins' and Kelvin's findings was the French astronomer and mathematician M. Delaunay (1868).[29] Although he maintained that nutation and precession of the globe gave no reliable indication about the thickness of its crust, he failed to make a strong case against Hopkins's results, as shown in a response by Kelvin in 1872. Similarly, the chemical geologist David Forbes argued against the concept of a solid Earth on the basis of the steep geothermal gradient, which he took as an indication of "a sea of melted rock" at relatively shallow depth.[30] Yet Forbes was fully aware of the effect of pressure on the melting point: "As it must be remembered, however, that at this depth the substance of the earth would be exposed to the pressure of the superincumbent mass, and as it has been demonstrated by experiment that many substances become more refractory, i.e. require a greater heat to melt them or keep them in the molten state when exposed to pressure." Citing experiments of both the German chemist William Bunsen and Hopkins on the effect of pressure on the melting point, Forbes concluded that "the solid rock crust of our earth cannot, at the utmost, be more than fifty miles in thickness." Beneath the crust was "the lava layer" with a thickness of approximately 650 km, which Forbes considered responsible for all volcanic and other igneous processes. The French geologist Elie de Beaumont (1871) also argued for a thin crust, about 45 km thick. Perhaps the most decisive contribution to this debate was Pratt's (1871) conclusion: "I think, therefore, that geologists must submit to the verdict that the crust of the earth is very thick, if not solid to the centre, and must be content with the idea that there are local seas of lava in the crust itself to account for volcanic phenomena."[31]

One geologist consistently opposed to the idea of a fluid Earth interior was George Julius Poulett Scrope, who in his 1862 publication *Volcanos, the character of their phenomena* spoke out vehemently against "an assumed sea of lava beneath the crust of the globe" and considered the Earth to contain only pockets or "vesicles" of liquid matter here and there in the interior.[32]

Further studies by Kelvin in 1876 led to revisions of Hopkins's earlier calculations, strengthening the concept of a thick outer shell of the Earth and prompting Kelvin to state: "This conclusion is absolutely decisive against the geological hypothesis of a thin rigid shell full of liquid." Perhaps Kelvin's best known work in geophysics was his application of the cooling rate of a spherical body to the study of the age of the Earth.[33] De Buffon in France had attempted a similar calculation in 1749 and Joseph Fourier applied his heat transfer equation to the solution of this problem in 1820, arriving at an age of the planet of 200 million years. Kelvin considered the Sun a cooling star that had illuminated the Earth for no more than 100 to 500 million years. He also believed that the temperatures and rates of heat dissipation in the early solar system were formerly much greater, and that geologic processes, such as volcanism, were once much more vigorous and must be slowing down because of gradual loss of heat from the planet. He envisaged the Earth as a heat machine that is continuously losing energy as heat is conducted through its crust and into the atmosphere. The consequences seemed obvious: "Within a finite period of time must have been, and within a finite period of time to come the earth must again be, unfit for the habitation of man as at present constituted, unless operations have been, or are to be performed, which are impossible under the laws to which the known operations going on at present in the material world are subject."

For earthlings contemplating the internal heat of the Earth it was only natural that they would cast their eyes to the Sun in the search of an explanation. Immanuel Kant had considered the Sun as a flaming body, not a mass of molten or glowing matter. In the context of nineteenth-century prenuclear-age science, it was clear that the Sun could not remain hot forever—it would eventually run out of fuel. And the German scientist Hermann von Helmholtz had pointed out that if the Sun were burning like a giant campfire, it would have run out of fuel in a mere thousand years. Instead he proposed gravitational contraction as the Sun's source of energy, with the settling of material to its center releasing gravitational potential energy in the form of heat (1847; 1856).[34] Compared to the campfire model, this was a highly productive and long-lived mechanism for the production of solar energy; it yielded an age of 20 to 40 million years for the sun. Even so, the solar thermal energy had to be decreasing with time.

Kelvin was greatly interested in heat conduction in the Earth and was aware of the rough estimates of the thermal gradient in the crust, as determined in mines. He was not impressed with the qualitative approach of his contemporary geologists and strongly advocated measurements of geophysical parameters as a means of furthering our understanding of the Earth.[35] In particular, he promoted the idea of a survey of the temperature gradient in the Earth's crust to answer fundamental scientific questions about our planet: "We must send out and bore under the African deserts. The whole earth must be made subject to a geothermic survey." By the end of the nineteenth century, Kelvin had estimated the Earth's thermal gradient as 1°C per 27.8 m.

That heat flows from hot to cold bodies is an illustration of one of the fundamental laws of nature. This principle of the dissipation of energy led to the Second Law of Thermodynamics, a result of the work of the French scholar Sadi Carnot on steam engines that described the conversion of heat to mechanical work.[36] Kelvin used this principle to determine the age of the Earth. He proposed that it first cooled by convection from a completely molten state as heat was rapidly transferred to the surface by upwelling hot currents from the interior. As it began to solidify, convection ceased, and subsequent cooling was by the slow process of conduction of heat upward through the solid crust. From his calculations of outward heat flow, measurements of the geothermal gradient in bore holes and mines, and certain assumptions of an initial temperature of the globe of 3900°C, he concluded in 1862 that the time since the crust first formed on a molten Earth was about 100 million years. Recognizing the uncertainties in his assumptions, Kelvin estimated that the Earth's age could be anywhere between 25 to 400 million years. Geologists, on the other hand, had estimated a vastly greater age for the Earth on the basis of their estimates of the time required to form all the strata that make up the geologic record. Geologists and biologists argued that even 400 million years was inconceivable for the observed duration of geologic and evolutionary processes; they also argued, on Uniformitarianism principles, against Kelvin's claim that geologic processes had been more vigorous in the past. Writing in 1862, Kelvin remarked that geologists had essentially ignored the principles of thermodynamics in favor of Uniformitarianism. However, his calculations, based on contemporary physics and precise experimentation, seemed so overwhelmingly logical that most geologists gradually capitulated to his erroneous view of the Earth as a very young planet. By 1899 Kelvin had shortened his estimate of the age of the Earth to a mere 20 to 40 million years. His heat flow calculations had been greatly refined as a result of the experimental work of Carl Barus (1856–1935) of the U.S. Geological Survey, who supplied new figures on thermal conductivity that gave a new age of the Earth's crust of 24 million years.[37]

The discovery of radioactivity a few years later was soon to provide another limitless supply of heat, rendering Kelvin's conclusions about the age of the Earth invalid. As pointed out by Frank Richter, however, it was not only Kelvin's ignorance of radioactivity as a source of heat that threw his calculations off, but rather that he did not take into account the full effects of thermal convection.[38] It is thermal convection that has transported heat from the interior of the planet toward the surface, thus cooling it much faster than through conduction alone. This importance of convection as an agent of heat transfer in the Earth—and its implication for a much greater age of the planet than proposed by Kelvin—had been already realized in 1895, but generally was ignored by geologists.[39]

The theoretical analysis of the internal structure of the Earth was carried further by the English geophysicist George Howard Darwin (1845–1912), the second son of Charles Darwin.[40] He evaluated the viscosity of the Earth's interior on the basis of observations of tidal motions of the Earth's crust and its resistance to the gravitational pull of the Moon and in 1878 concluded that "the results are fully as hostile to the idea of any great mobility of the interior of the earth as is that of of Sir W. Thomson [Kelvin]." To resist the tidal force of the moon, the Earth's interior must possess a viscosity greater than that of cold pitch. At that time, the viscosity of cold pitch seemed very high indeed; fifty years later geophysicists came to the realization that a substance with such viscosity can indeed flow within the Earth, given sufficient time. Kelvin was very pleased with Darwin's efforts, which essentially supported his own earlier conclusions of a solid Earth interior, and praised the young man as "the chief guardian of the Rigidity of the Earth." From his studies, Darwin concluded "that no very considerable portion of the earth can even distantly approach the fluid condition."[41] He realized, however, that somehow he had to account for magma within the Earth as a source of volcanism, which he explained as a result of the "existence of liquid vesicles in the interior, or by the melting of solid matter, existing at high temperature and pressure, at points where dimunition of pressure occurs."[42]

Darwin's arguments for a largely solid Earth, based on gravitation studies, were soon supported by direct observation of the transmission of earthquake waves through the Earth. In the second half of the nineteenth century, seismology, or the science of the study of earthquake waves, was developed. In England, Robert Mallet built seismographs, measured seismic velocities, and compiled the first modern catalogue of recorded earthquakes. And the British physicist John Milne (1850–1913), working at the Imperial College of Engineering in Tokyo, took advantage of the high frequency of earthquakes in Japan, where he "had the opportunity to record an earthquake every week." By the 1890s Milne and other British physicists working in Japan had established that earthquake waves pass through most of the interior of the Earth. This was a particularly important finding; because fluids cannot transmit transverse waves (called S waves), it follows that the Earth's interior must be solid. By 1899 it was clear that the Earth is an elastic solid throughout the greater part of its mass and is capable of transmitting seismic waves. As instrumentation and analysis of seismographs became more precise, it became possible to define the geometry of bodies within the Earth with differing densities. Thus, Richard Dixon Oldham (1858–1936) first showed in 1906 that the Earth has a largely molten core, based on his observation of greatly reduced velocities of seismic waves traveling through the Earth's central region. Because this deep region did not transmit S waves, it had to be molten. At about the same time, it was confirmed that the region between the crust and the core, which became known as the Earth's mantle, transmitted both S and P waves, and was therefore regarded as solid, with a thickness of 2900 km. By the 1920s seismology had also shown that there could be no continuous layer of melt underneath the crust. The view of the Earth's structure at the turn of the century was defined as a series of concentric solid shells or layers, with an outermost 30 to 40-km thick low-density layer of granitic crust (sial) making up the continents, underlain by a very thick layer (sima or the mantle) of high-density ultrabasic rock with the composition of peridotite, and finally the central core. Those who were searching for a molten region in the Earth as a source of magmas were now faced with a geophysical picture of a largely solid interior. A magma source in the molten core seemed impossible, at a depth of over 2900 km below the surface.

14

Melting by Decrease in Pressure

Let this tremendous pressure be locally relaxed by the fissuring and elevation of the resisting crust, as has evidently happened along the lines of mountain chains, and portions of the solid mass below must immediately pass into the fluid state—giving rise to the igneous intrusive rocks so generally found within and beneath them.

George Julius Poulett Scrope, 1868

The discovery that the Earth beneath the crust was essentially solid completely eliminated an "undercrust fire-sea" that Dana and others had proposed as a source of the magmas that feed the Earth's volcanoes. Scholars, now faced with the formidable task of showing how hot magma could be derived from within a solid Earth, found in the science of thermodynamics a partial solution to the problem. Thermodynamics is the science of the power of heat to do work and of the dissipation of that power. The conversion of motion to heat had been known ever since humans rubbed sticks together to light a fire. The conversion of heat into work, such as during expansion of heated gases, is a principle that was well-known to the ancients. That air expanded on heating was recognized in Hellenistic times, and by 1600 Galileo was aware of the potential uses of this phenomenon. But it was probably thanks to the practical applications of heat in the steam engine that the science of thermodynamics developed during the Industrial Revolution, attracting some of the best intellects of nineteenth-century Europe. Studies of the conversion of heat to mechanical energy became the research topic of highest priority for engineers and scientists because it had profound economic significance. These studies were also important in comprehending the nature of heat in the Earth as well as the process that brings about the melting of rocks to produce magma.

By the early years of the eighteenth century it had been fully appreciated that pressure influences the temperature at which substances will undergo a change of phase—such as from a liquid to a gas or a solid to a liquid—during melting. The German physicist Gabriel Daniel Fahrenheit (1686–1736) had demonstrated in 1724, for example, that atmospheric pressure has a systematic effect on the boiling point of water, causing boiling to occur at lower temperatures if pressure is decreased. And the investigations of John Ellicott (1736), who devised an instrument to measure the degree of expansion of metals due to heat, showed the effect of heat on the expansion of solids. In the case of the melting of rocks, it was realized, first by intuition, then from experiment, and finally from theory, that the temperature of melting increases with depth—that is, that the melting curve has a positive slope.

Count Rumford is often regarded as one of the founders of thermodynamics. He was born in America as Benjamin Thompson (1752–1814) and was to become one of the greatest applied scientists of all time. His most famous experiment is the cannon-boring experiment, which he carried out in Munich in 1802, in which he made the first determination of the mechanical equivalent of heat and showed that heat is a "mode of motion."[1] In this experiment he observed with surprise the great amount of heat produced by the action of the boring tool on the gun and demonstrated that the heat flow from a body is inexhaustible, which discredited the caloric theory of Lavoisier in favor of a mechanical explanation. Rumford also demonstrated that heat is not caloric or material by showing that it had no weight.

During the first half of the nineteenth century a handful of French, German, and English scientists laid the foundations for the new science of thermodynamics. In a few key publications, fundamental equations were established that describe the relationship between the temperature, heat content, pressure, and volume of gases and liquids. In 1822 Joseph Fourier published his work on the dissipation of heat, and in 1824 the former French army officer Sadi Carnot was the first to formulate the theory, simultaneously describing the process of conversion of heat and mechanical energy and conservation of energy in a slender volume with the title *Reflexions sur la puissance motrice de feu*. This led to the Second Law of Thermodynamics describing the interconversion of heat and mechanical work. Although Carnot focused his ideas on the development of steam power and industrialization in France, he did point out that even earthquakes and volcanic eruptions are the result of heat.

Because of a growing interest in thermodynamics, melting experiments with various chemicals and natural substances were carried out at a range of temperatures and pressures, establishing that the melting point of many substances increases with pressure. The possible effects of pressure on melting of rocks in the Earth had been qualitatively appreciated quite early in the development of geologic thought. Perhaps the first recognition of this fundamental effect was made in 1802 by John Playfair in his *Illustrations of the Huttonian Theory of the Earth*.[2] He pointed out that just as a change in pressure affects the boiling point of water, so would melting in the Earth be influenced by the great pressure exerted by the overlying rocks. In his discussion in the chapter "Compression in the Mineral Regions," Playfair wrote:

"The fact, of water boiling at a lower temperature under a less compression, is sufficient to justify the supposition, that bodies may be made by pressure to endure extreme heat, without the dissipation of their parts, that is, without evaporation or combustion." Although his main concern was demonstrating that high pressure could result in the formation of marble in contact with an intrusion of hot magma, he was also clearly aware of the effect of pressure on the melting point of rocks. Again, in 1805, he addressed the effect of pressure, stating, in connection with James Hutton's opposition to melting experiments on rocks, that "Dr. Hutton himself...considered the fact of the liquification of mineral substances by heat as so completely established, that it affords a full proof of the fusibility of those substances having been increased by the compression which they endured in the bowels of the earth. In his view of the matter, no other proof seemed necessary."

Playfair and Hutton both maintained that a fire within the Earth was impossible because combustion cannot occur in regions of "such compression" where the air vital to burning is absent. They conceived of a heat distinct from fire, favoring instead the theory of primordial heat in the Earth as a cause of melting and a source of volcanism. How melting was brought about or that melting of a hot and solid rock at depth in the Earth could occur by a decrease in pressure was not stated. Of this heat source, Playfair wrote (1802): "There is nothing *chimerical* in supposing that nature has the means of producing heat, even in a very great degree, without the assistance of fuel or of vital air. Friction is a source of heat, unlimited, for what we know, in its extent, and so perhaps are other operations, both chemical and mechanical; nor are either combustible substances, or vital air, concerned in the heat thus produced. So also the heat of the sun's rays in the focus of a burning glass, the most intense that is known, is independent of the substances just mentioned."

The proof of the effects of pressure on mineral stability was provided by the high-pressure and high-temperature experiments of Playfair and Hutton's friend, Sir James Hall, in Edinburgh. Hall had carried out more than 500 experiments between 1798 and 1805 on the stability of calcium carbonate.[3] Joseph Black, Hall's teacher, had demonstrated in 1756 that upon heating in the atmosphere, calcium carbonate breaks down and gives off "fixed air" or carbon dioxide gas, but Hall successfully showed that carbonate did not decompose during heating under high pressure. When he simulated the high pressures within the Earth by heating the calcium carbonate in tightly sealed gun-barrels, he found that it recrystallized to marble.

The next to address the importance of pressure in volcanic systems was George Poulett Scrope, born George Julius Thomson, the son of a London merchant. After a brief stint at Oxford University in 1815, he entered Cambridge University in 1816, where his geologic education benefited from the teaching of Adam Sedgwick. In Cambridge he enrolled as George Poulett Thomson, adopting his mother's aristocratic name. He developed a deep fascination for geology and this interest was further stimulated by family excursions to Italy, where he began a lifelong fascination with volcanic activity.[4] According to Scrope, he first became interested in "the phenomena of Volcanos in the winter of 1818, when I passed within sight of Vesuvius, then in permanent eruption. In the course of the next year I visited Etna and the Lipari Isles, particularly Stromboli, where I spent some days. After having examined the phenomena of these active volcanos, I turned my attention to the extinct vents of Italy and France; passed six months in Auvergne, the Velay, and Vivaris, continually occupied in geological researches, and afterwards revisited Vesuvius (which I reached just in time to witness the stupendous eruption of October 1822), examined the Ponza Isles, and all the different volcanic districts of Italy from thence to the Euganean Hills, returning to England by those of the Rhine, the North of Germany and the Eifel."[5] He returned to the Italian volcanoes in 1819, and to the volcanoes of Auvergne in 1821 for six months of fieldwork, then again to Italy in 1822. He later wrote of his excursions: "We can never forget the luxury of geologizing in the extinct craters of the Elysian fields [Phlegraean Fields], with Virgil in one hand and a hammer in the other."

In 1821 Thomson married the heiress of the aristocratic Scrope family of Castle Combe in Wiltshire; on this occasion he changed his name again, to George Poulett Scrope. For a time, he continued his volcanological studies with vigor and originality and his writings earned him membership in the Geological Society of London in 1824; he served as a secretary of the Society with his friend Charles Lyell in 1825. In 1826 he became a Fellow of the Royal Society. At this time he published in rapid succession two important works: *Considerations On Volcanos* (1825) and *The Memoir on the Geology of Central France* (1827).

The young Scrope then became, perhaps, a victim of his own financial security following his marriage.[6] At any rate, his geologic research went into eclipse from about 1831 to 1856. As a country gentleman he had taken a strong interest in economics as well as local history. In 1832 he became a member of Parliament, practiced liberal politics, and worked toward the improvement of the condition of the working classes until his resignation in 1867. He wrote at this time a number of pamphlets on economics, local history, and the Irish famine, and his penchant for issuing numerous but short publications earned him the sobriquet "Pamphlet" Scrope.

By 1856 Scrope's interest in volcanism had been rekindled and he published several papers related to this topic until the year of his death. These were not, however, based on new fieldwork, but were instead reviews and comments on the studies of other scholars. No longer able to travel abroad because of declining health and failing sight, he continued to financially support some young geologists in their fieldwork in distant lands. His last publication was related to the volcanic eruptions of the Askja caldera and the Sveinagja fissure in Iceland, in 1875.[7]

In *Considerations on Volcanos* Scrope discussed the combined effects of pressure and temperature on separation or exsolution of water from magma:

> Let us examine the effect of any variations in the circumstances of temperature and pressure on a substance of this nature, viz. a solid crystalline rock containing a certain proportion of water intimately combined between the crystalline molecules, and already possessing a high temperature, more than

> equal to the vaporization of the water under the pressure of the atmosphere alone. It is obvious that either the augmentation of temperature, the pressure remaining fixed, or the diminution of the pressure, the temperature being constant, must alike have the effect, sooner or later, of producing the commencement of ebullition in the confined water.... Having thus far considered the effect of an increase of temperature, or a diminution of pressure, on a mass of lava under such circumstances, let us examine what will follow from the reverse; namely, an increase of pressure, or a diminution of temperature. Upon the solid lava, it is clear, no corresponding change will be produced; but every diminution of temperature, or increase of pressure, on a mass, or a part of the mass, liquified in the manner stated above, must occasion the condensation of a part of the vapour which produces its liquidity, and so far tend to effect its *reconsolidation*.[8]

Although he was clearly aware of the potential of decompression melting, he did not, however, propose in his 1825 work that melting in the Earth was due to pressure release. Instead, he envisioned a solid interior, where that elusive material, fluid heat (caloric) rises from the core into the overlying crystalline rocks and brings about their melting. He had a qualitative grasp of the fundamental principles governing the effects of temperature and pressure on melting, but his view of melting and generation of magma was still clouded by the caloric theory of the nature of heat.

Later, Scrope explicitly addressed the relationship between temperature, pressure, and the melting point of rocks in his critical review of Lyell's *Principles of Geology* in 1835 in which he pointed out that at certain depths in the Earth, the pressures due to the influence of gravity were sufficiently great to "counteract the liquifying tendency of the temperature, however great, and retain the nucleus in a solid form."[9] This was the first statement on the *slope* of the melting curve of the Earth: rocks will melt at higher temperature at depth than near the surface. In his book *Volcanos, the character of their phenomena* he also noted "the antagonistic influences of high temperature in resisting the solidification of mineral matter, and of great pressure in promoting it," and thus implied that melting can be brought about by a decrease in pressure.[10]

Unlike many of his contemporaries, Scrope was convinced that the interior of the Earth was largely solid due to the effect of high pressure on the melting of rocks, and that only pockets or "vesicles" of molten material remained here and there at depth.[11] Such vesicles, he proposed, "sometimes solidified by increase of pressure, then again, perhaps, liquified by its diminution, or increments of caloric reaching them laterally or from beneath by conduction." The mechanism of melting was in his view clearly due to decompression: "Let this tremendous pressure be locally relaxed by the Fissuring and elevation of the resisting crust, as has evidently happened along the lines of mountain chains, and portions of the solid mass below must immediately pass into the fluid state—giving rise to the igneous intrusive rocks so generally found within and beneath them." When he had first proposed decompression some forty years earlier as the principal cause of melting, the reception was generally unfavorable: "For entertaining this view, I was at that early period subjected to much ridicule, and my arguments generally disregarded." By 1868, however, he maintained that his views on melting were accepted by "many—perhaps the majority of geologists."

The Melting Curve of the Earth

When William Hopkins began his studies on the question of the fluidity of the Earth, he was immediately confronted with the nature of the distribution of heat and the melting curve of the planet's interior.[12] Having no direct experimental evidence to build on, Hopkins, like Playfair and Scrope before him, based his opinion mostly on intuition, but was probably aware of Playfair's proposals and especially those of his contemporary Scrope on the effect of pressure on melting, although he does not refer to either of them. Speaking of the relationship between depth (i.e., pressure) and temperature, Hopkins wrote: "To estimate this tendency under the joint influence of these causes, it would be necessary, in the first place, to know the law according to which the temperature increases in descending from the surface to the centre, while the mass is cooling by circulation; *and secondly, the influence of the temperature in resisting solidification, as compared with that of the pressure in promoting it*" (emphasis mine).[13] Hopkins continued: "These, however, are points on which we possess at present little or no experimental evidence." Although he proposed that pressure increases the melting point of rocks, he did not, however, take the credit for this as an original idea, mentioning in a footnote that Simeon-Denis Poisson (1781–1840) was the first to suggest it. Playfair and Scrope, for unknown reasons, were not given deserved credit.

Four years earlier, Poisson had suggested in his work *Théorie mathématique de la chaleur* (1835) and also in a paper in 1837 that, during the formation of the Earth, the outer portions of the hot and molten sphere would sink toward the center, where high pressure would maintain all matter in a solid form. The excessively high pressure in the deep interior of the Earth, he proposed, would lead to solidification of rock material at much higher temperatures than at the low pressures near the surface; he even postulated that the solidification of the Earth had commenced at the center and progressed outward, to the surface.[14]

Documents of the Royal Society show that one of those who reviewed Hopkins' 1839 manuscript was George Biddell Airy, the Astronomer Royal, who apparently took issue with Hopkins's claim that increasing pressure leads to a higher melting point.[15] In a paper in 1842, Hopkins again pointed out that knowledge of the effects of pressure on the temperature of fusion within the Earth was crucial for a proper understanding of the planet's interior, proposing that "with the aid of a proper series of experiments on this point, a direct method of arriving at an approximation to the thickness of the crust of the globe, or rather to its least limit, might be easily explained." Yet, unable to inspire collaborators—or raise money—these experiments would not take place until nine years later.[16]

At about the time that Hopkins was publishing his work, a fundamental thermodynamic equation was first obtained by the French scientist Benoit-Pierre-Emile Clapeyron (1799–1864) that describes the change in volume when a solid substance melts. Clapeyron kept the work of Sadi Carnot alive when he published *The Motive Power of Heat* in 1834 because he gave analytical form to Carnot's ideas. Clapeyron demonstrated that all bodies in nature change their volume with change in temperature or pressure, and that this applies equally to liquids, gases, and solids, including the Earth. It would seem that the four principal thermodynamic properties—volume, pressure, temperature, and heat content—are interrelated; if any two properties are taken as independent variables, then the other two can be considered as functions of them. In 1850 the German scientist Rudolph Clausius, in an article *On the Motive Power of Heat*, further developed Clapeyron's theory. The combined efforts of these two scientists, who are generally regarded as among the founders of thermodynamics, led to the formulation of the Clausius-Clapeyron equation, important because it quantifies the relationships between temperature, volume, and pressure. Clausius also showed experimentally in 1850 that if a given quantity of a substance has a greater volume when melted to a liquid than in its solid or crystalline state, an increase in pressure would lead to an increase in the temperature of melting of this solid. His, Clapeyron's, and Kelvin's work led to the development of a number of thermodynamic principles that describe the direct relation between the change in volume of a substance on melting or crystallization and the effects of pressure on the temperature on melting or crystallization. The Clausius-Clapeyron equation, which may be written as

$$\frac{dT}{dP} = \frac{T\Delta V}{\Delta H}$$

considers two phases of the same chemical substance, for example, a liquid and solid (magma and rock), in equilibrium with one another at temperature T and pressure P. By supplying heat slowly to the system, one phase changes reversibly into another (i.e. brings about melting), with the system remaining at equilibrium. In the equation, the fraction dT/dP represents the rate of variation of the melting point T with change in pressure P. The value ΔV represents the volume change on melting and is generally positive (i.e., the specific volume of the solid is smaller than the volume of the corresponding liquid). The equation expresses the variation of pressure and temperature for a system in equilibrium.

At about the same time (1849), the effects of pressure on crystallization and melting of ice were being theoretically evaluated by the physicist James Thomson (1822–92), Kelvin's older brother. Thomson demonstrated that changes in pressure have a direct effect on the freezing point of water and derived a formula that gives the freezing point in terms of the pressure.[17] Thomson argued that since water expands on freezing (contrary to the volume increase upon melting of most other substances, including rocks), it could be argued that the freezing of water involved expansion and the production of work. His results indicated that the melting point of solids, other than ice, became more elevated when subjected to pressure, in agreement with Carnot's and Clapeyron's theory. After all, the slope of the melting curve dT/dP was determined by the sign of the volume change of the substance in question on solidification or freezing. Kelvin's experimental verification of his brother's calculations led to the development of the Kelvin scale of absolute temperature.[18]

It is not difficult to see why most materials expand on melting and why their solid phase is denser and sinks through the liquid. In a solid the atoms or molecules are closely packed, often in a dense crystalline structure. In crystals the atoms are arranged in an extremely regular and close-packed lattice network, with the atoms vibrating about their average position even though they are not free to wander within the rigid structure. When heat is added, the atomic vibration and the temperature of the solid increase. If enough heat is added, the vibration energy will overcome the binding forces of the crystal lattice and the atoms will escape from their regular sites and begin to wander. Eventually, the rigid structure will dissolve and the material melts. In the molten state, the atoms or molecules are no longer closely packed but move in endless agitation as a loose mass, one that is much less compact than the preexisting solid. The behavior of materials is, however, also affected by pressure. If a liquid just above the melting point is subject to high pressure, the atoms of the liquid are again forced into close proximity by the pressure and resume their places in a crystal lattice. The material solidifies, even though the temperature of the material remains constant. Thus, the melting point of most substances increases with pressure; so does the boiling point, for the same reasons.

It was shown early in the eighteenth century, in melting and crystallization experiments, that the specific volume of a volcanic rock such as basalt is lower than that of the corresponding magma. Many geologists who studied columnar basalt correctly interpreted its structure as evidence of the contraction of magma on cooling and solidifying to rock. During cooling and solidification, the material shrunk in volume, forming roughly hexagonal columns separated by a pattern of contraction joints. Thus, it was clear that the term ΔV in the Clausius-Clapeyron equation was positive; consequently, the equation predicts that the pressure-temperature melting curve (dT/dP) of the source rock of basaltic magma has a positive slope in the Earth—that is the temperature of melting increases with pressure.

This interpretation of the columnar basalt was consistent with the experimental results of the German chemist Gustav Bischof in 1837, who carried out one of the first measurements on the contraction of basalt and other volcanic rocks on fusion and showed that rocks contract on solidification from magma.[19] He concluded that a melt of granite rock contracted about 25% on solidification, the volcanic rock trachyte about 18%, and a basaltic melt 11%. Bischof also carried out detailed investigations on the temperature of soils, water in mines, and rock outcrops in order to study heat distribution within the Earth. He convincingly showed that temperature increases with depth in mines, although the rate may vary from place to place, and that the rate of temperature increase varies also with depth. Bischof considered two possible causes of volcanism: either chemical action or very high original temperatures within the Earth. Chemical action, he was to conclude, would be insufficient to account for the heat. Instead, he proposed that the source rocks of magmas must be in a state of near-fusion at depth. But because the melting point of different rocks is so variable, it would follow that at any given level, some rocks could be in a state of fusion, surrounded by solid or crystalline rocks capable of withstanding higher temperatures. His experiments had shown that the oxides of sodium and potassium were fluxes, which promoted melting, and consequently he proposed that the most easily fusible rocks were those rich in alkalis and containing feldspar, mica, and leucite. Bischof proposed that the mean melting point of lava is 1268°C, and used this figure to estimate the depth of melting in the Earth. If temperature increases with depth in the manner predicted from surface measurements, then in his opinion the Earth must melt at a depth of about 38 to 42 km beneath Vesuvius and Etna. Bischof supposed that steam, derived from sea water and heated by volcanism, is the power by which the magma is raised from great depths to erupt at the surface. He evaluated the expansion force of steam at various depths and temperatures, and concluded that, at magmatic temperatures, the force of steam would be sufficient to propel magma upward.

Another German chemist, Robert Wilhelm Eberhard Bunsen (1811–99), was one of the first to consider the effect of pressure on the generation of magmas and to experiment on the relation between pressure and melting point of substances.[20] The laboratory facilities available in Bunsen's time (1850) permitted only modest pressures to be achieved, roughly the equivalent of the pressure at a depth of 1 mile in the ocean. He therefore performed his experiments on materials with relatively low melting points, such as ambergris and paraffin, and extrapolated these results to the high pressures and temperatures prevailing in the deep Earth. His work showed that a pressure increase of only 100 atmospheres increased the melting point of these substances by several degrees centigrade (Figure 14.1).

Bunsen concluded that rocks could melt at a certain temperature at low pressure, but would crystallize or remain totally solid at a higher pressure, even at the same temperature. Similarly, at a constant pressure, different types of rock may generate different magmas: "And when the change in the melting point is different in different rock formations subject to the same change in pressure [as his experiments had shown for ambergris and paraffin], so can the effect of pressure change the order of magma generation under certain circumstances and influence the chemical composition of the magma." Bunsen was thus the first to propose that

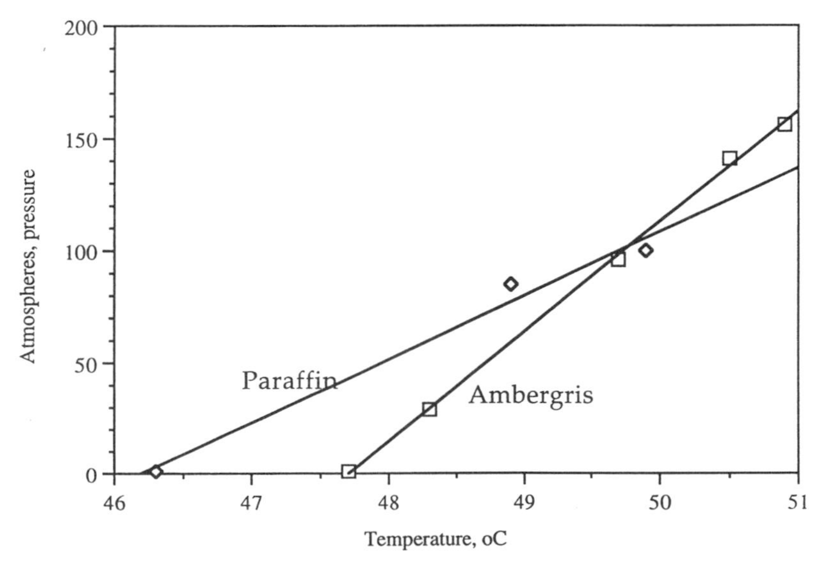

14.1. In 1850, the German chemist and geologist Robert Bunsen found that an increase in pressure had a profound effect on the melting point of both paraffin and ambergris, as shown. The results of such experiments were the qualitative basis for proposing that the melting point of rocks also increased in the same manner with pressure or depth in the Earth.

different magma types could be generated by the variable effects of pressure on the melting curve, and also at constant pressure, because of the different melting temperature of various rock types. Bunsen also suggested that pressure influences the phase relations (the interdependence of the composition of coexisting minerals and melts) of minerals on melting and crystallization. Bunsen's co-worker during an expedition to study the volcanic rocks of Iceland was the German geologist Sartorius von Waltershausen, who also evaluated the relationship between pressure and the melting point of substances and concluded that, for silicate rocks, the melting point increases about 1°C per 100 atm of pressure.[21]

Other scientists recognized the effect of pressure on melting as an important area of research. William Hopkins, who had finally secured a grant of £250 from the Royal Society to continue research, began an experiment in 1851 with James Joule, Kelvin, and the engineer William Fairbairn on the effects of pressure on the solidification of the Earth's interior. For this experiment, which took place in Manchester, Fairbairn supplied a large lever apparatus for generating high pressures, initially up to 400 atm, but later as high as 5400, equivalent to the pressure at approximately 15 km depth in the Earth. The great mechanical facilities of the industrial city of Manchester were thus put to good use for science.[22] Initially the experiments were on substances with low melting points, such as beeswax and spermaceti. Progress appears to have been slow—perhaps the apparatus needed refining—but in 1853 Hopkins reported: "At present our experiments have been restricted to a few substances, and those of easy fusibility; but I believe our apparatus to be now so complete for a considerable range of temperature, that we shall have no difficulty in obtaining further results. Those already obtained indicate an increase in the temperature of fusion proportional to the pressure to which the fused mass is subjected. In employing a pressure of about 13,000 lbs. to the square inch on bleached wax, the increase in the temperature of fusion was not less than 30° Fahr."[23] In 1855 further reports came, establishing that the melting temperature of such substances as sulfur and stearine increased with pressure, in the range within his experimental capabilitites—that is, up to 800 atm and 158°C. Hopkins's health failed in the year he reported on these results and consequently his research on the effects of pressure on the melting curve came to an end (Figure 14.2).

While it was now clear that pressure increased the melting point of most substances, the major question remained unresolved: how to extrapolate from these experiments to the high pressures and temperatures in regions where magmas are actually generated. Because the experiments fully supported the theory embedded in the Clausius-Clapeyron equation and in the thermodynamics of James and William Thomson, the slope of the melting curve in the Earth could then, in principle, be deduced from the equation and from knowledge of the volume change of volcanic rocks on melting. But, without high-temperature melting and solidification experiments on rocks, the question of the sign of the volume change on melting and crystallization of magmas—did rock expand or contract on melting?—created considerable controversy among scientists. Mallet had found that magma decreased in volume by about 6%, but others claimed the opposite.[24] The English volcanologist John Judd (1881) was also

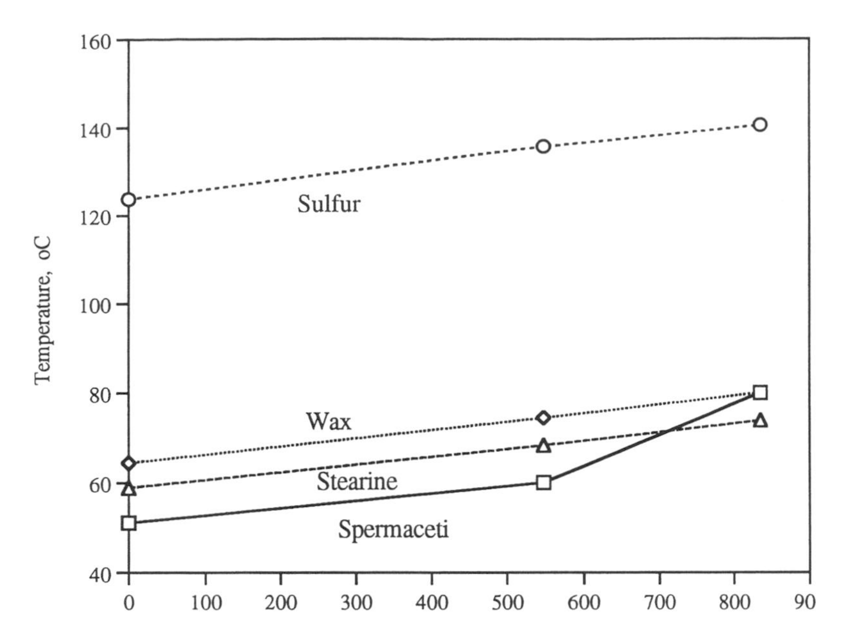

14.2. In 1855 William Hopkins presented results on the effect of pressure on the melting point of various substances obtained experimentally with his collaborators J. J. Joule and William Fairbairn, using the relatively low-melting compounds of sulfur, spermaceti, wax, and stearine, but working at pressures and temperatures much higher than in Bunsen's experiments in 1850. The graph shows the increase in the melting point of these substances with increasing pressure.

aware that the melting point of most solids is raised with pressure, and that rocks deep in the Earth, which are at temperatures far above their normal melting point at the surface, could be considered in a "potentially liquid condition" because they would begin to melt with any decrease of pressure. Yet he did not regard this a major factor in the generation of magmas, and was inclined to the chemical theory for the origin of heat in the Earth.[25]

In 1878 the influential American geologist Clarence King began to address the nature of volcanic phenomena in the context of the new science of thermodynamics. Discarding the hypothesis of an Earth with a molten interior, he proposed, following Hopkins and Kelvin, that the rate of increase of pressure with depth is greater than the rate of increase of temperature. Thus, the actual temperature gradient of the Earth would always be significantly below the melting curve, and the interior of the planet would be unmelted and rigid. He believed that local relief of pressure leads to melting in the Earth and suggested that the pressure release could occur with the erosion and removal of overlying crustal rocks: "So that the isolated lakes of fused matter which seem to be necessary to fulfill the known geological conditions may be the direct result of erosion." This seemed perfectly logical at the time, but, if true, the rate of erosion of a given area must occur at a higher rate than that of heat conduction from the rising hot rock. Geologic evidence did not support this theory on two counts. First, many mountainous regions, where erosion is highest, show no signs of volcanism, and, second, the rate of pressure relief due to erosion is so slow that heat lost through conduction would prevent melting at depth.[26]

By 1889 a number of volcanologists had fully recognized that pressure relief was necessary to bring about melting in the Earth. Among them was the British geologist Logan Lobley who suggested that "should, however, at any part of a sufficiently highly-heated interior region, pressure be removed, or sufficiently lessened, the hindrance to fusion would cease to exist and solid rock would become at once fluid lava." Pressure relief could come about by "the local elevation of the uppermost crust, or by fracturing or giving way of rocks below volcanic areas" where there would be found "more or less permanent reservoirs or underground lakes of fluid lava." Lobley stated: "Rock fusion, resulting from simple relief of vertical pressure in subterranean regions where the heat is sufficient to fuse rocks under surface conditions, would not be limited in lateral extension if there is an open vent, and surface depression and derangements with consequent lava outputs on a scale far transcending any terrestrial catastrophes that have occurred would result."[27]

The foremost volcanologist in Britain at the end of the nineteenth century was Archibald Geikie (1835–1924). After many decades of work on volcanic rock formations in Scotland, Ireland, France, and the United States, he published an important book titled *The Ancient Volcanoes of Great Britain*.[28] Although Geikie devoted his work primarily to the study of volcanic landforms and surface processes, he also expressed his views on the generation of magmas. He accepted that the Earth's interior was entirely solid, despite the rapid increase in temperature with depth: "At a depth of a few miles, every known substance must be hotter than its melting point at the surface. But at the great pressures within the earth, actual liquification is no doubt prevented, and the nucleus remains solid, though at a temperature at which, but for the pressure, it would be like so much molten iron." Geikie then succinctly pointed out that melting in the Earth could be brought about by decompression and proposed a geologic mechanism:

> Any cause which will diminish the pressure may allow the intensely hot material within the globe to pass into the liquid state. There is one known cause which will bring about this result. The downward increment of temperature proves that our planet is continually losing heat. As the outer crust is comparatively cool, and does not become sensibly hotter by the uprise of heat from within, the hot nucleus must cool faster than the crust is doing. Now cooling involves contraction. The hot interior is contracting faster that the cooler shell which encloses it, and that shell is thus forced to subside. In its descent it has to adjust itself to a constantly diminishing diameter. It can only do so by plication [folding] or by rupture.

Geikie's vision of interior thermal contraction leading to deformation of the Earth's crust was probably greatly influenced by the writings of Charles Darwin and the American geologist James Dana.[29] Although in some areas the crust is folded, in others it rises up in sharp folds, which he proposed as a way that pressure immediately below could be relieved—promoting melting and generation of magmas. This view was similar to the collision of two ice flows on a river, with one riding up over the other. The fundamental error in his model was the assumption that the Earth's crust has great strength. On length scales of hundreds of kilometers, the crust is weak and cannot support its own weight or tolerate large stresses, and, when subject to great lateral pressure, it will deform or break up. Geikie fully understood decompression melting, but, like his contemporaries, was groping for a viable geologic mechanism to produce the pressure release.

While geologists continued to speculate, a handful of geophysicists were conducting experiments related to the problem. The fundamental measurements on the thermodynamic properties of basalts were made possible near the end of the nineteenth century. The Reverend Osmond Fisher promoted experiments to determine the specific heat of basalt up to 1200°C; these experiments also showed that a large amount of heat is absorped during melting, in the range of 800 to 900°C.[30] The most important measurements, however, were made by Carl Barus (1856–1935), a physicist with the U.S. Geological Survey.[31] Barus began pioneering experiments in the Survey's laboratory in 1883, under the guidance of its director, Clarence King. Starting in 1889, Barus carried out a series of studies related to the question of origin of magmas, at pressures of 2000 kg/cm^2 (equivalent to 6 km in the Earth). He determined that basalt rock shows a sharp increase in volume of about 6 to 7% when it melts at 1100°C. He also determined the latent heat of fusion for basalt, which enabled him to apply the Clausius-Clapeyron equation to determine the effect of pressure on its melting point. He concluded that the term dT/dP in the equation was equivalent to 0.025 for basalt—that is, the melting point increases at a rate of 2.5°C for every 100 atm increase in pressure (or about 300 m increase in depth).[32] Having provided this key evidence, he went no further and concluded his short note with the cryptic statement: "The immediate bearing of all of this on Mr. Clarence King's geological hypothesis is now ripe for enunciation." In 1893 Carl Barus determined the melting curve of basalt as a function of pressure and King then made use of the Barus melting curve for basalt, extrapolating it to elevated pressure, to attempt to place constraints on the geothermal gradient in the Earth.[33] King evaluated a series of temperature gradients for the Earth, which were calculated by Kelvin's method for an Earth that has cooled by conduction, following Fourier's law of the dissipation of heat. Given the fundamental assumption of a solid Earth, as deduced from its rigidity to tidal forces, the geothermal gradient was constrained to fall below the basalt melting curve throughout the Earth and King proposed that a gradient nearly tangential to the melting curve was likely to reflect the true temperature distribution (Figure 14.3).

In Europe, research on this problem was also advancing, but opinions greatly differed on the shape of the melting curve of the Earth. In 1900 the Swedish scientist S. A. Arrhenius (1859–1927) reviewed these conflicting opinions. The geophysical evidence, Arrhenius pointed out, favors the view that the interior of the Earth must be largely solid. Although its temperature is likely to be extremely high, the rocks there are subject to such enormous pressures that their melting point is elevated and they would not melt. In his opinion, the melting curve flattened out with pressure and intersected the geothermal gradient at a relatively shallow depth. Consequently, the Earth was in a fiery and fluid condition at a depth of several tens of kilometers below the surface. Adopting a geothermal gradient of 30°C per km, Arrhenius arrived at a temperature of about 1200°C at a depth of 40 km. There, he proposed, magma generation occurred in a molten zone of 100 to 200 km thickness. At a depth of 300 km he envisioned the temperature to be so high that it exceeded the critical temperature of all known materials, with the magma consequently in a gaseous form.[34] Few were willing, however, to consider the presence of such a widespread molten zone in the Earth, but instead

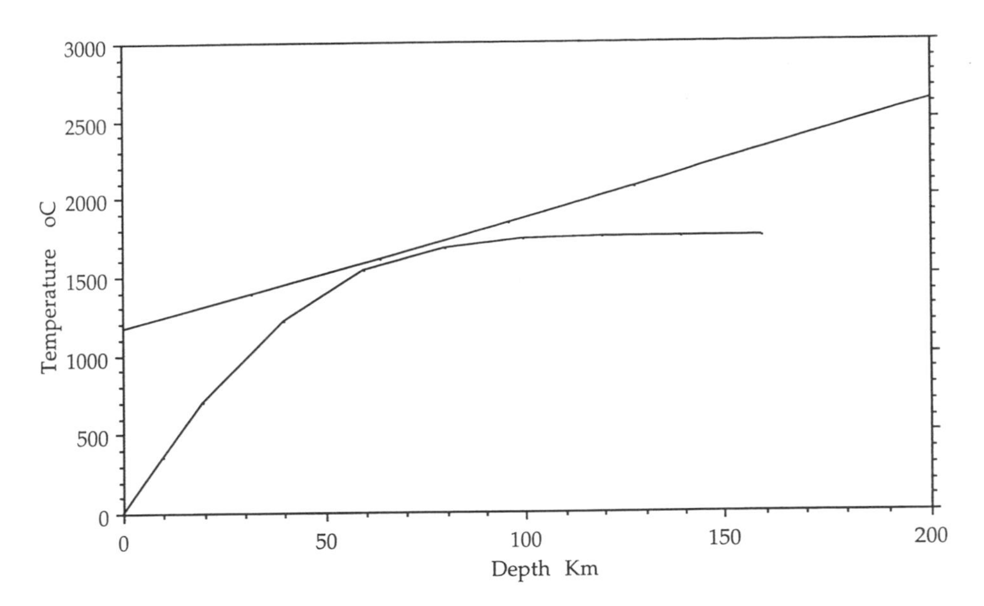

14.3. The curved line shows the geothermal gradient of the Earth, according to the American geologist Clarence King in 1893. King estimated the geothermal gradient on the basis of observations made in bore holes for shallow regions; he also adopted Kelvin's calculations, based on the cooling of a sphere. He then adopted the melting curve for basalt at elevated pressures (straight line), as determined by his co-worker Carl Barus, and used this constraint to level off the geothermal gradient so that it would not intersect the melting curve at any depth. King thus drew the geothermal gradient based on the belief that the Earth was essentially solid in this depth range. Melting would occur where the geothermal gradient intersects the basalt melting curve, or around 60 km and 1600°C.

regarded the production of magma as a consequence of local relief of pressure. Thus, in 1909 the British geologist Alfred Harker concluded:

> It appears then, that we must seek the immediate cause of igneous action, not in the generation of heat, but chiefly in *relief of pressure* in certain deep-seated parts of the crust where solid and molten rock are approximately in thermal equilibrium. We are thus led by an independent line of reasoning to the principle already enunciated, which connects igneous action primarily with crustal stresses, and so secondarily with crust-movements...at a sufficient depth, such conditions of temperature prevail that solid and liquid rock are in approximate thermal equilibrium. Any local relief of pressure within that region, connected with a redistribution of stress in the crust, must then give rise to melting.

These statements were the most incisive expression of the fundamental principle of melting in the Earth to date; what was still lacking, however, was a full understanding of the mechanism that brought about the relief of pressure.[35]

By 1914 the American physicist P. W. Bridgman (1882–1961) at Harvard University had constructed an experimental apparatus capable of attaining pressures up to 13,000 kg/cm^2, equivalent to the pressure at approximately 4 km depth in the Earth's crust. His results showed that Carl Barus's thermodynamic predictions were valid: the temperature of melting continued to rise with increasing pressure. Bridgman also demonstrated that the volume change on melting of metals and some salts remained positive, even under very high pressure.[36] In 1923 the internally heated pressure vessel was invented, making it possible to carry out melting experiments on rocks at pressures up to 10 kbar (equivalent to a depth of 30 km) and temperatures up to 1600°C. Using of the new techniques, the American geologist Reginald Aldworth Daly was able to determine in 1933 that the melting point of diopside (an important mineral in the Earth's mantle) increases about 4.6°C per kilometer of depth in the Earth's crust.

Thus, in the early part of the twentieth century, the most attractive physical mechanism to bring about melting in the Earth was recognized as decompression, fully supported by geophysical experiments. But geologists were by no means ready to embrace this as a solution to the origin of magmas because a viable geologic process that could bring about a pressure decrease was not known. Daly states: "Local relief of pressure is hopelessly inadequate and need not be further discussed."[37] Also, according to the American geologist James Shand, this pressure effect on the melting of silicates was "a quantity scarcely large enough to have any important petrological consequences." He considered the only significant pressure effect that of boiling vapor phases in magmas on low-pressure decompression, such as what occurred during the ascent of magma.[38]

The progress made in understanding melting and the internal constitution of the Earth by the first part of the twentieth century was truly profound, but it had created a paradox. All the evidence was in favor of a solid interior—or an exceedingly thick crust at any rate—yet there was a need to account for the magmas erupted from volcanoes. Experimentalists had discovered a process by which melting could happen simply by decompression, but geologists were unable to find an acceptable decompression mechanism to produce this melting.

15

Radioactive Heat and Convection

If there be heat in the centre of the globe, it must have the properties of heat and none other. I ask not how the Heat originally was lodged in that situation, for the origin of all things is obscure; but I ask why, in the countless succession of ages which the Huttonian requires, the Heat has not passed away by conduction, and if it has passed away, by what other heat it has been replaced?

George Greenough, 1834

In concluding his analysis of the Sun's heat and the secular cooling of the Earth, Lord Kelvin made the following cautionary statement in 1862: "As for the future, we may say, with equal certainty, that inhabitants of the earth cannot continue to enjoy the light and heat essential to their life, for many million years longer, unless sources now unknown to us are prepared in the great storehouse of creation." He added, however, "I do not say there may not be laws we have not discovered." Only thirty-four years later a new source of heat, radioactivity, was discovered—one that would revise Kelvin's findings drastically, supplying a virtually inexhaustible heat source for the Earth's interior.

Wilhelm Conrad Röntgen (1845–1923) may be credited with the discovery of radioactivity. In 1895, working in a darkened laboratory with a vacuum tube covered in black paper, he noticed that when he passed a current through the tube, a "peculiar black line" appeared on a piece of barium-coated paper on a table next to the tube—a shadow that could have been caused only by radiation other than light. These rays were produced when cathode rays, or accelerated electrons, impinged on the barium-coated paper; Röntgen called these X rays. For his discovery he was awarded the Nobel Prize in 1901. By 1896 the French physicist Antoine-Henri Becquerel (1852–1908) had observed that the uranium compound potassium uranyl sulfate spontaneously emitted rays that were energetic enough to blacken a photographic plate, even when it was wrapped in black paper. A few years later Marie and Pierre Curie succeeded in isolating a tiny amount of this radiating element from many tons of the uranium ore called pitchblende; the element was a grey metal rather like barium in its properties—and they gave it the appropriate name: radium.

The new radioactivity caused a great stir among European scientists, who quickly applied it to an understanding of the Earth. In 1903 Pierre Curie had also discovered that radioactive materials emit heat and that the heat emission from radium was so great it could easily be detected with an ordinary thermometer. In 1903 the Irish physicist John Joly was the first to point out in the journal *Nature* that the abundance and distribution of radium and other radioactive elements may determine terrestrial heat. C. H. Liebenow then calculated that a global concentration of only sixty parts per billion of radium, distributed evenly throughout the Earth, was sufficient to account for the observed heat conduction through the crust of the planet.[1] Suddenly, geologists were freed from the dogma of a primordial or original heat in the Earth—heat that had been steadily declining since the planet's creation. The British physicists Ernest Rutherford (1871–1937) and Frederick Soddy (1877–1956) also began to study the effects of the heat generated by radioactivity and showed that a lump of radium produces enough heat to melt its weight of ice in one hour, and would continue, apparently, to emit heat for an infinitely long time. They proposed in 1904 that radioactivity may be responsible for heat in the Earth: "The temperature gradient observed in the earth may be due to the heat liberated by the radioactive matter distributed throughout it." This led Rutherford to conclude in 1904 in a classic understatement that "the time during which the Earth had been at a temperature capable of supporting the presence of animal and vegetable life may be very much longer than the estimate made by Lord Kelvin from other data." Rutherford and Soddy established that the heat was almost entirely due to the emission of alpha particles, released in the explosive disintegration of the elementary atoms. The alpha particles are positively charged nucleii of helium atoms, ejected at very high velocity. Their collisions with other atoms are the main source of heat generated during the process we commonly refer to as radioactive decay.

In 1906 R. J. Strutt (later Lord Rayleigh) published a classic paper on the radioactivity of rocks, based on his detection of radium in a great variety of common rocks from many parts of the Earth. He showed that the amount of helium in minerals, produced by the decay of uranium, was much greater than what could have been produced given the age of the Earth Kelvin proposed; he indicated the age of the Earth could be as much as 2.4 billion years. By 1908 Joly had presented his ideas on radioactive heating of the Earth, suggesting a nearly eternal heat source.

Following these discoveries it became clear that the cooling and contraction within the Earth is dependent on the half-life of uranium, the principal nuclear fuel of the planet. The most important of the radioactive, or unstable, elements in the Earth is uranium, which decays into radium and lead, and thorium, which also decays into lead. The important characteristic of a radioactive element is its half-life—the time it takes for one-half of the original atomic nucleii of a radioactive element in a given specimen of material to break up or decay. Some elements have half-lives of a fraction of a second; others are millions of years. Those found in the Earth (mainly uranium, thorium, and potassium) typically have very long half-lives; the more unstable and short-lived elements have decayed and completely disappeared long ago. The half-life of uranium is about 5 billion years, whereas thorium has a half-life of 14 billion years. Since the age of the Earth is about 4.5 billion years, it is evident that the initial inventory of these radioactive elements has been only partly depleted; they will continue to provide a kind of "fire" in the interior of the Earth for many millions of years to come.

In 1910 the British geologist Arthur Holmes (1890–1965) joined Lord Rayleigh's laboratory to study the natural radioactivity of rocks and thus began a career that would make a major contribution to the understanding of both heat distribution and the melting of the Earth by convective flow of solid rocks toward the surface. By 1915 Holmes had calculated a temperature profile for the Earth (temperature distribution in the interior) based on radioactive generation of heat.[2] Later the British geophysicist Sir Harold Jeffreys arrived at a thermal gradient significantly shallower (lower temperature at comparable depth) than Holmes had proposed (Figure 15.1).

The thermal gradient reflects two factors: the generation of heat in the deep Earth, and heat loss by convection and conduction through the crust. The flow of heat from the interior through the Earth's crust is determined by measuring the temperature gradient in near-

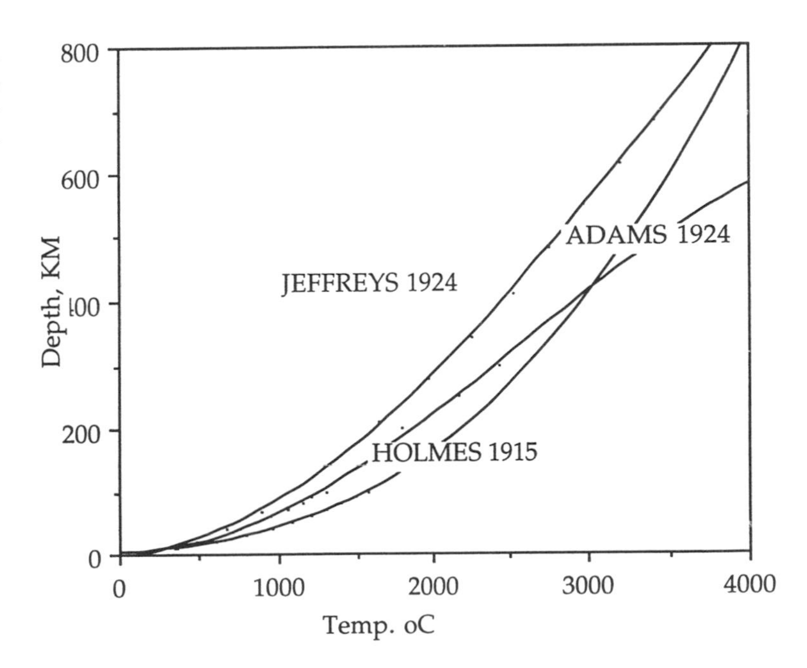

15.1. Three opinions on the geothermal gradient in the Earth, according to Arthur Holmes (1915), Harold Jeffreys (1924), and Leason H. Adams (1924).

surface rocks and how well they conduct heat. The heat flow out of the Earth is about 40 million megawatts. Compare this to the electric generating power capacity of the United States, which is about 400,000 megawatts, only about 1% of the global heat flow. The Earth also loses heat through volcanic eruptions, which amounts to a heat loss of about 2 million megawatts. Volcanic heat flow averages only about 5% of the total flux of heat from the Earth's interior, but is very high locally, as in the glacier-covered Grimsvötn caldera in Iceland that melts 0.5 km^3 of the overlying ice sheet each year, an energy output of 5000 megawatts, and the Yellowstone caldera in Wyoming, which produces about 4200 megawatts. In Yellowstone the high heat flow is, of course, manifest in boiling groundwater, producing spectacular geysers and hot springs.

Four main sources of energy in the Earth must account for the observed heat flow. One is the original heat, its heat content since the time of its formation. The others are gravitational energy, kinetic energy associated with the tidal pull of the Sun and the Moon, and radioactivity. Simple reasoning, however, shows that radiogenic heat production can account for nearly all the heat flow.[3] The important heat-producing radioactive elements in the Earth are the isotopes of uranium, thorium, and potassium. Concentrations of these elements in the Earth have been estimated, on the basis of chemical composition of mantle rocks and meteorites, as 18 parts per billion of uranium, 65 parts per billion of thorium, and about 800 parts per million of potassium. The rate of heat generation from their radioactive decay amounts to about 38 million megawatts, or very close to the observed heat flow value of 40 million megawatts. But how is this great flux of heat transferred from the deep Earth toward the surface?

Over time, physicists have considered three modes of heat transfer: conduction, radiation, and convection. If the Earth's heat loss were due solely to conduction, then the interior would not yet "know" that the outside is cold. Rocks are excellent thermal insulators, and the values of thermal conductivity in rocks of the deep Earth are so low that only the planet's outermost 400 km would have lost significant heat by conduction during its 4.5 billion-year lifetime.[4] As early as 1928 Holmes proposed that convection was a much more effective mechanism for transporting heat from depth to the surface. In 1797 Count Rumford was the first to propose convective flow in fluids as a means of dissipating heat, and by 1839 William Hopkins had advocated convection in the Earth's interior, an idea later applied to geological problems by the Reverend Osmond Fisher.[5] In *Physics of the Earth's Crust* (1881) Fisher proposed that convection currents existed in the Earth's fluid interior, rising under the ocean and descending beneath the continents—and that they furnished a "means of obtaining those local increments of temperature which in some form or another appear to be needful in order to explain volcanic phenomena." In 1886 he also advanced the idea that these currents transported ocean crust toward continental crustal areas, thus formulating a crude but totally new form of the plate tectonics cycle. Kelvin, Fisher's contemporary, was quite familiar with convection and had in fact discussed convection in connection with the thermal structure of the atmosphere, but did not consider it important in terms of temperature distribution within the Earth. Instead, he viewed heat loss from the Earth solely as the result of conduction.

After radioactivity was discovered as an internal heat source, it became even more important to establish the principal means of heat loss from the planet. A. J. Bull proposed in 1921 that differential radioactive heating set up currents within the Earth, but it was Arthur Holmes who discovered the relationship between convection in the Earth and the movement of the great crustal plates. In a paper published in 1928, "Radioactivity and Earth Movements," Holmes proposed that the excess heat is discharged from the Earth by the circulation of material in the "*substratum*," or the Earth's mantle, forming thermal convection currents. Convection, and in turn continental drift, was seen as a mechanism to rid the Earth of the great heat generated by the radioactive natural reactor. He considered this substratum a solid peridotite rock, which could flow like a fluid at the scale of time and distance that characterizes a deep Earth process—about 5 cm a year. Peridotite is a strikingly green and beautiful rock that is composed dominantly of the mineral olivine; peridot is its gem quality. It is relatively rare at the Earth's surface, but fragments of it are thrown out of volcanoes. To most geologists it seemed likely that peridotite makes up the mantle—about two-thirds of our planet. Holmes' scheme included downward currents below geosynclines, the great sedimentary troughs that we recognize today as subduction zones, and ascending currents below midoceanic "swells," where "a discharge of a great deal of heat" occurs due to upwelling of peridotite mantle, decompression, and melting. Holmes's views in 1928 were amazingly modern and laid the foundations for the concept of plate tectonics, although Earth scientists have not generally recognized this.

Holmes presented all his ideas in his classic text *Principles of Geology*, first published in 1944. By 1964 he had shown that the temperature gradient through the Earth was everywhere far below the melting temperatures of the common igneous rocks, with a gap of a few hundred degrees. He proposed convection currents in the mantle as a mechanism to bridge that gap, by the convective flow of hot but rigid material to shallower levels in the Earth where the mantle material was transported across the melting curve, leading to the generation of magma.

The classic problem of determining whether convection is possible was solved by Lord Rayleigh. The occurrence of convection in the Earth depends on the depth of the layer in question, the temperature gradient, and the viscosity of the material; the ratio of these factors is known as the Rayleigh number. The critical value for convection to occur is a Rayleigh number of 1000 or higher. Whether convection occurs depends on the viscosity of the Earth's mantle. If viscosity is too great with respect to the temperature gradients, convection may be inhibited because the retarding viscous forces win over the buoyancy forces associated with the decreasing density of hotter material.

The Petrology of Volcanic Rocks

The recognition of peridotite as the rock that makes up the mantle was made possible by the development of crystallography and petrology. Perhaps Pliny expressed the earliest appreciation for crystals in his *Natural History* in the first century: "Why crystal is generated in hexagonal form it is difficult to assign a reason; and the more so since, while its faces are smoother than any art could make them, the pyramidal points are not all of the same kind," Nicolas Steno in 1669 made the great intellectual leap that created the science of crystallography when he established that while the sides of the hexagonal quartz crystals may vary, the angles between them do not. By 1783 the French geologist Romé de Lisle had described some 500 regular forms of crystals, but it was the French scholar René Just Haüy who made mathematical deductions on the symmetry, form, and dimensions of crystals. Haüy's *Traite de mineralogie* (1801) is a landmark in the classification and identification of rock types as well as in the definition of a system of rock names based on their mineralogy.[6] Minerals are crystalline substances, and a few of these—such as feldspar, pyroxene, olivine, hornblende, and magnetite—are the rock-forming minerals that make up volcanic rocks.

The study of the textures, mineralogy, and chemical composition of rocks is termed *petrology*, derived from the Greek word *petros*, or rock. Among the earliest observations on the mineral content or petrography of volcanic rocks were Leopold von Buch's studies in 1799 on the volcanic rocks near Rome. In his paper "On the Formation of Leucite" von Buch concluded that the crystals of leucite had formed while the lava was still fluid, and discounted any hypothesis to the effect that they had been precipitated from an aqueous solution and later incorporated into the magma. Von Buch's view was, of course, anti-Neptunist, although he still was uncertain about the volcanic origin of basalt.

The science of petrology began to develop in 1811, when J. Pinkerton published *Petrology: A Treatise on Rocks*. Basalts and other volcanic rocks are composed of exceedingly fine-grained crystals, which are invisible to the naked eye or even through a magnifying glass. The pioneer geologists were therefore unable to investigate the internal structure of these rocks and identify their tiny crystal components. The microscopic study of volcanic rocks was first carried out on crushed rock powder by Pierre Louis Cordier (1777–1861) in 1815, who crushed basalts and other volcanic rocks to a fine powder, separated the various particles by a flotation process, and examined them by microscopic and chemical tests. He concluded that ancient basalts were very much like modern volcanic rocks in texture, mineralogy, and other characteristics—an important step in solving the Neptunist versus Volcanist controversy over the origin of basalts.[7] Cordier's studies also showed that basalts and some lavas were essentially alike, whereas various sedimentary rocks were quite different in their physical, mineralogical, and chemical properties. He established that the minerals in volcanic rocks were either dominantly feldspar or pyroxene and consequently proposed a classification of these rocks as belonging either to the *feldspathic* or *pyroxenic* division. Then in 1844 Charles Darwin, who on his voyage on the H.M.S. *Beagle* had made an astonishing number of observations on volcanism, published *Geological Observations on the Volcanic Islands*, in which he classified igneous rocks as either acid or basic in composition, referring to the content of silica or "silicic acid." The principal division into acid (rhyolites) and basic (basalts) is still fundamental to the classification of volcanic rocks today.

Rock can be sliced and polished to a plate so thin that it is transparent to light and easily examined under the microscope. The art of making thin sections may have begun with the sectioning of fossil wood, but this method was soon applied to rocks. Their minerals could be identified and their textures studied in detail. By 1829 the polarizing microscope had been invented by William Nicol (1768–1851) and was used widely for the study of crystals. Henry Clifton Sorby (1826–1908) was the first geologist to make thin sections of volcanic and other igneous rocks for microscopic study.

Another crucial step in the study of volcanic rocks was the development of their chemical analysis. Plutonists and Neptunists alike agreed that the fundamental basis for the petrology of rocks was their chemical composition. However, the chemical analysis of rocks is not a simple matter because they are exceedingly resistant to most chemical solvents and even to strong acids, except hydrofluoric acid. Studies of the chemical composition of volcanic rocks date back at least as far as the work of Abbé Lazzaro Spallanzani (1794) who reported on them in his monumental study of the volcanic rocks of Italy in terms of a weight percentage of the five major oxides of silica, alumina, lime, magnesia, and iron. He discussed the obsidian lava of Lipari and was the first to recognize that this rock type is natural glass.[8] As mentioned previously, in 1805 the Scottish chemist Robert Kennedy reported on his chemical analyses of *whinstone* or basaltic rocks. Having shown that basalt from the columns of Staffa in Scotland was virtually insoluble in aqueous and acid solutions, he was able to dissolve the rock by melting rock powder with caustic potash at high temperature in a silver crucible. He then demonstrated that it contained 48 wt% silica, and 16% alumina, 9% lime, and 16% iron oxide. Water and other volatile components constituted about 5%. About 6% was not accounted for and Kennedy suspected this was in part "a saline substance" and, after an elaborate analytical routine, showed that it was about 4% soda. He thus accounted for 99% of the chemical constituents of the rock—a great feat at that time.[9]

When the early geologists began to accumulate data on volcanic rocks, they discovered that the chemical composition of the rocks, even those from the same volcano, varied greatly. Confronted with the problem of how to account for this diversity, George Scrope proposed in 1825 that all types of igneous rocks were formed from a single parent magma that gave rise to a variety of types through a process he referred to as *liquation* and *differentiation* before crystallization. Charles Darwin, however, laid the foundation for the process of "magmatic differentiation" through his discovery of crystal settling due to density differences between crystals and magmatic liquid. A parental magma can change its chemical composition due to the formation of crystals in the magma, which sink to the bottom of the magma reservoir, and in this manner certain chemical components are subtracted, resulting in the formation of another magma type. During his exploration of the volcanic Galapagos Archipelago in September 1835, Darwin studied a young basaltic lava flow on the west side of James Island (now San Salvador). Here he noted that the crystals of feldspar were much more abundant in the base of the lava than in the upper part and deduced "that the crystals sink from their weight."[10] Von Buch had earlier (1818) noted a similar concentration of crystals in the lower part of obsidian lava flows on Tenerife in the Atlantic, interpreting this as the settling of crystals in the lava during and after flow, but had not appreciated the broader implications. Darwin, on the other hand, considered this observation as "throwing light on the separation of the trachytic and basaltic series of lavas" (trachyte is a magma type with high silica, whereas basalt contains low silica). He realized that once early formed crystals sink, the remaining magma would be chemically different from the initial liquid. This idea is central to the concepts of differentiation and fractional crystallization, which account for much of the diversity of the chemical composition of volcanic rocks erupted at a given volcanic center. In support of his hypothesis, Darwin pointed out "that where both trachytic and basaltic streams have proceeded from the same orifice, the trachytic streams have generally been first erupted, owing, as we must suppose, to the molten lava of this series having accumulated in the upper parts of the volcanic focus."

Another and radically different process was developed by Robert Bunsen in 1851 to account for the chemical range of volcanic volcanics, after his discovery of two magmas in Iceland, with radically different chemical composition.[11] After the great eruption of Hekla volcano in 1845, the King of Denmark asked Robert Bunsen to carry out geologic studies in Iceland (Figure 15.2). Bunsen was struck by the abundance of two volcanic rock types: almost white, yellow, and multicolored silica-rich volcanic rocks and the dark colored basaltic rocks. He carried out chemical analyses of these rocks and reported their content of the oxides of

15.2. The German chemist Robert Wilhelm Bunsen (1811–99) pioneered the chemical analysis of volcanic rocks. From a late nineteenth-century painting by J. Marx.

15.3. When Robert Bunsen and Sartorius von Waltershausen visited Iceland in 1846 they were struck by the spectacle of hot springs and geysers as manifestations of internal heat in the Earth, such as those shown in this 1813 engraving of the Great Geyser, with Hekla volcano in the background (author's collection).

silica, alumina, iron, calcium, magnesium, potassium, and sodium. A few volcanic rocks were of intermediate composition and he proposed that they were the result of mixing of the silcic and basic magmas. He showed mathematically that mixing could account for the intermediate volcanic rock types. Bunsen's contribution was to put forward the first viable hypothesis that could account for the diversity of the chemical composition of volcanic rocks by the mixing of two primary magma, basalt and rhyolite. He did not, however, offer an explanation for the origin of these primary magmas. Traveling with Bunsen in Iceland was Sartorius von Waltershausen, who earlier had made studies of the volcanic rocks from Mount Etna (Figure 15.3).[12] He was the first to discuss the distribution and occurrence of the trace elements (chemical constituents present in very low abundance) as well as the major oxides in volcanic rocks, and listed twenty-three trace elements.

In the eighteenth century the volcanic fluid had become known as magma, derived from the Greek word that refers to a plastic mass, such as a paste of a solid or liquid matter. The term magma was first used in pharmacy as in "magnesia magma," or milk of magnesia, and the word passed from pharmacy into chemistry to represent a pasty or semifluid mixture. From chemistry it passed into petrology to replace "subterraneous lava." What was the nature of this fluid inside the Earth? Bunsen stressed that magmas are nothing more than a high-temperature solution of different silicates: "No chemist would think of assuming that a solution ceases to be a solution when it is heated to 200, 300 or 400 degrees, or when it reaches a temperature at which it begins to glow, or to be a molten fluid."[13] This was a fundamental breakthrough in thinking about crystallization from magmas. Bunsen further recognized that minerals crystallized from magmas at a much lower temperature than the melting point of the pure minerals. If magmas are true solutions, then they should be able to mix, as indeed Bunsen had proposed on the basis of his Iceland work. This idea was applied further to volcanology by David Forbes in 1871, who recognized mixed magma eruptions when he states "that from the same volcanic orifice, and during the same eruption, lavas of two totally different classes may be emitted, viz. the light acid or trachytic lava, analogous to the granites, felsites etc., of the oldest periods, and the heavy basic pyroxenic lava, all but identical with the dark basaltic or trappean rocks."[14] Not all volcanologists were convinced that magmas were solutions, however. Scrope wrote in 1872 that "lava is not a homogenous molecular liquid, but a magma, or compound of crystalline or granular particles to which a certain mobility is given by an interstitial fluid."

Another process proposed in the nineteenth century to account for the diversity of magma types was Soret's principle. Soret had demonstrated experimentally that if a tube is filled with a salt solution and the two ends are maintained at different temperature, the fluid at the colder end will become richer in salt; thus, the specific gravity of the fluid in the tube is no longer uniform but increases from the warmer to the colder part. T. C. Bonney showed that Soret's principle could not explain the process of differentiation of magmas.[15] If the more basic (denser) components of magma became concentrated in its colder regions, the specific gravity of the mass would increase in the opposite direction from the temperature—and the exterior of a deep-seated magma body or an intrusion would be more basic in composition than its acidic interior. Similarly, the upper layers of the Earth should be more basic than the deeper regions. This would lead to a gravitational instability (rapid sinking of the crust), which caused Bonney to reject the hypothesis. On the other hand, Bonney considered crystal settling a more attractive process for explaining the differentiation of magmas.

During the latter part of the nineteenth century the chemical analysis of volcanic rocks became commonplace because of rapid developments of chemical techniques. When the American geologist Frank W. Clarke presented the first data on the relative abundance of the elements in the Earth's crust in 1889, seventy elements were already known.[16] He worked with a compilation of 880 analyses of rocks from the continents to obtain a mean crustal composition for the nine rock-forming elements. Leading petrologists of this age believed that most igneous rocks were derived from basaltic magma by a process of differentiation, but no one researcher contributed more to the understanding of the processes that account for the chemical diversity of igneous rocks than the American petrologist Norman L. Bowen.[17] From high-temperature experiments, Bowen discovered that magma could greatly change in compostion as it cooled if the early formed crystals were effectively separated from the liquid, thus providing proof for the fractional crystallization process first proposed by Darwin. Bowen was a strong advocate for fractional crystallization as the main process responsible for differentiation and considered basaltic magma the primary or parent magma of all volcanic rocks. This new understanding of magmatic differentiation was made possible by Bunsen's recognition that magmas were true solutions, Darwin's concept that settling of crystals resulted in a change in the composition of the liquid, and Bowen's systematic experiments on crystallization of a primary basaltic liquid. But how, then, was this primary basaltic liquid formed?

16

The Source of Magmas

Although convection might lead to melting, it cannot be shown that it does, and the problem of the generation of magma remains as baffling as ever.

Francis Turner and John Verhoogen, 1960

Once it had been established that magmas were generated by the melting of solid rock within the Earth, the nature and chemical composition of the source rock became the focus of much research. Throughout the first half of the twentieth century, two principal hypotheses were debated on the nature of the source. Because the principal type of magma erupted was basalt, it appeared logical to some that the source region was basaltic in composition, possibly the high-pressure form of basalt known as eclogite. The basaltic magma would then be derived by wholesale melting of the source. The other view was that the magma was derived by partial melting of a different type of rock named peridotite. It had long been noted that lumps, or *xenoliths* (Greek for foreign stone), of both peridotite and eclogite were ejected from some volcanoes. Since they were brought up in the magma, it seemed reasonable that these rock types might be present in the source region of the basaltic magmas.

The long process of discovery of peridotite as the magma source within the Earth may have begun with an observation made by Charles Darwin. In February 1832 Darwin landed on the tiny pinnacles known as St. Paul's Rocks, situated almost exactly on the equator in the Atlantic Ocean.[1] During his voyage, Darwin had noted that virtually all the mid-ocean islands were volcanic; St. Paul's Rocks, however, were clearly an exception. He noted an abundance of a dark green mineral (serpentine) in the rocks, but was at a loss as to the origin of this small island: "It is composed of rocks, unlike any which I have met with, and which I cannot characterise by any name." Mystery continued to surround the remote St. Paul's Rocks long after Darwin's visit. Because of their unique position near the axis of the great undersea mountain chain known as the Mid-Atlantic Ridge and their unusual composition, the oceanographer A. Renard speculated in 1882 that they might represent the pinnacles of the lost continent of Atlantis.[2] We now know that St. Paul's Rocks are made up of peridotite, a beautiful bright green, crystalline rock that contains an abundance of the mineral olivine. Later work has shown that St. Paul's Rocks are a sliver of the Earth's mantle that has been thrust up through the crust of the underlying ocean floor.[3] Darwin in fact had walked on rocks that make up the interior of the Earth and represent the source of magmas.

As early as 1879 it was suggested that peridotite was the source of basaltic magmas and by 1916 Arthur Holmes proposed that magma was generated by the melting of peridotite at a depth of several hundred kilometers due to radioactive heating of the Earth.[4] Both Joly and Holmes had pointed out that all rocks contain some radioactive materials that can generate heat in the rocks and contribute to melting and the origin of volcanism. Joly's theory included a cyclic heat release, where radioactivity produces heat in the Earth's mantle and heat accumulates faster than it can escape by conduction. Eventually enough radioactive heat is produced to melt basalt, which expands and moves upward, with basaltic magma escaping through fissures and producing lava flows at the surface.

Holmes modified Joly's hypothesis and proposed radioactive heat generation in peridotite rocks at depth in the Earth. As mentioned in the previous chapter, he suggested that one very important result of this heat generation is convective circulation in the mantle, due to the upwelling of hotter and less-dense material. The circulation system he proposed within the Earth is similar to that of the atmospheric circulation, with slow ascending currents in equatorial regions that turn polewards as they rise and approach the crust. This would produce a depressed zone, an oceanic belt, in the equatorial regions, caused by drag on the continental blocks by the flow (Figure 16.1).

At the beginning of the twentieth century, geophysical measurements were also yielding results in favor of a peridotite source. This was done by measuring the physical properties of a number of rock types under high pressure and temperature in the laboratory. By comparing these results with the speed of earthquake waves traveling through the deep Earth, it was shown that properties of the Earth's interior are comparable to those of peridotite and other ultrabasic rocks under great pressure in the laboratory. Geophysicists therefore also adopted a peridotite composition for the mantle.[5]

For many geologists, the ideas Bowen presented in his book *The Evolution of the Igneous Rocks* (1928) were the most influential on the origin of magmas.[6] Although he agreed that relief of pressure within the Earth could convert crystalline rock to liquid, he maintained that

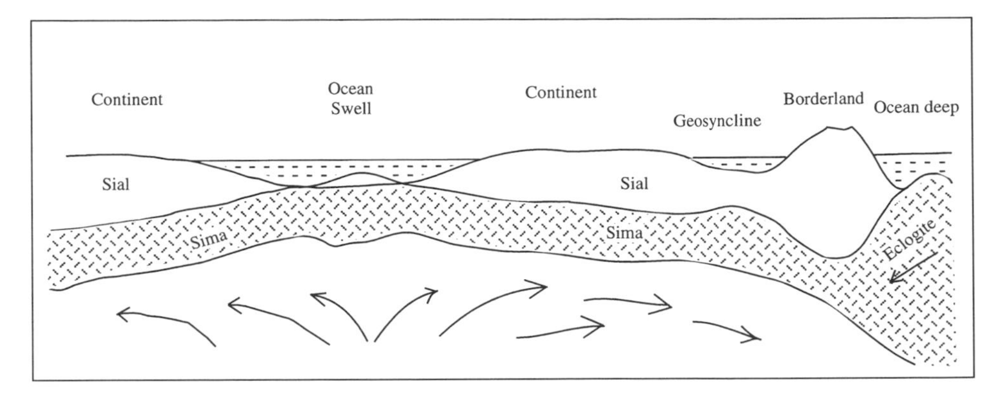

16.1. Arthur Holmes proposed that the mantle of the Earth beneath the crust was convecting and that volcanic activity occurred where convection cells diverged or converged. From G. Tyrrell (1931) after A. Holmes.

it would not bring about total melting of the region affected. He also pointed out that the only rock type that could give rise to basaltic magma by partial melting is peridotite and suggested that this occurs at depths of 75 to 100 km (47 to 63 mi). Arthur Holmes's ideas on convective flow in the mantle were published in the same year as Bowen's classic work, but Bowen seemed unaware of them. Having no plausible mechanism to bring about a pressure release, Bowen stated: "It seems necessary to leave open the question whether selective fusion takes place as a result of release of pressure or as a result of reheating." He therefore had considerable difficulty in developing a scenario of pressure-release melting in the peridotite mantle and the problem remained unsolved for another forty years. Only with the development of geodynamics and the acceptance of Holmes's theory of mantle convection could it be shown that a peridotite mantle can rise to regions of lower pressure, thus bringing about melting without the addition of heat.

The theory of partial melting of peridotite was still debated and considered on a weak foundation by many in the first half of the twentieth century. Another group of geologists adopted a totally different view and maintained that basaltic magma was derived from a deep melted layer or resulted from the complete melting of a basaltic layer at depth in the Earth. This view persisted until the 1960s. In 1936 Alfred Rittmann published the influential book *Vulkane und ihre Tätigkeit* (Volcanoes and Their Activity).[7] This work, like all later volcanology texts, was concerned primarily with descriptive surface aspects of volcanism—but it also attempted to account for the origin of magmas and melting in the Earth. All volcanic rocks, Rittmann proposed, were ultimately derived from a global layer of basaltic magma. He envisioned this layer at a depth of 35 km (22 mi) beneath the oceans and about 70 km (44 mi) beneath the continents. Rittmann's views were adopted by the Norwegian petrologist Tom F. W. Barth in 1951: "The fact that wherever the crust is deeply rifted, be it in continents, oceans, or geosynclines, basaltic magma is available and capable of invasion is a proof of the existence of a subcrustal basaltic magma stratum."[8] Barth was fully aware of the objections of geophysicists that such a layer could not transmit some seismic waves, contrary to observations, and proposed that the basaltic magma behaved like glass at the high pressures beneath the Earth's crust and was thus able to transmit these waves. "It is necessary, therefore, that the basaltic substratum is in a molten (non-crystalline) state at its subterranean locale. Otherwise it would not reach the surface with a homogeneous composition as we actually observe." In Barth's view, all that was required to initiate a volcanic eruption or an intrusion of magma into the Earth's crust was the opening of an abyssal fissure to the basaltic magma stratum at depth. These ideas of a molten substratum, which go back to the pre-Kelvin era of Gustav Bischof (1837), Amos Eaton (1830), Descartes (1644), and even as far as Empedocles in the fifth century B.C., seemed far-fetched to most geologists.

Unlike Rittmann and Barth, the American geologists Francis J. Turner and John Verhoogen accepted the seismic evidence for a solid character of the Earth's mantle in 1951 and considered it either composed of peridotite or eclogite.[9] From this it followed that primary basaltic magma had to be derived from a largely solid mantle, whose composition was other than basalt. The observed temperature increase with depth and heat generation by radioactive decay within the Earth could produce the required heat source. "The problem of the origin of basaltic magma is thus not so much that of finding adequate heat sources as that of explaining how relatively small amounts of heat may become locally concentrated to produce relatively small pockets of liquid in an otherwise crystalline mantle." In what seemed like a fleeting moment, Turner and Verhoogen considered the hypothesis that basaltic magma could be produced by the melting of a peridotite mantle due to pressure release—but then quickly rejected it: "This hypothesis of magma generation is possibly the most popular one, although it is unacceptable to the present writers. It is difficult indeed to see how pressure can effectively be reduced at such depths." They realized, of course, that melting could result if convective transport of mantle material occurred at shallower levels, as originally proposed by Arthur Holmes. This was clearly a viable mechanism so long as the deep mantle temperature was greater than the melting point of the mantle at shallower depth. Yet they were not ready to accept the existence of convection in the solid Earth, which led them to conclude: "Whether convection does occur in the mantle is not definitely known, nor is it known whether it could be effective in the upper part of the mantle where magmas are generated. Thus, although convection might lead to melting, it cannot be shown that it does, and the problem of the generation of magma remains as baffling as ever." This statement is a good reflection of the state of affairs regarding theories of magma generation at the middle of the twentieth century. Although great strides had been made in understanding the chemistry and mineralogy of volcanic rocks, the Earth scientists were at a loss to explain melting. This was due almost entirely to the static view taken of the Earth's interior at the time—it seemed inconceivable to most that the rigid and solid mantle could convect. However, the discoveries made on the ocean floor in the 1960s and the development of geodynamics were to change all that.

Although melting was poorly undestood at the time, knowledge of the mantle source region was advancing through studies of xenoliths and meteorites. Meteorites are fragments of the interiors of other planets that have disintegrated and fallen to Earth. Their study has helped us piece together the nature of the Earth's interior. An even more important line of evidence about the composition of the deep Earth comes from xenoliths. They are rock fragments, a few grams to kilos in size, ejected from volcanoes during eruption. A great number of xenoliths are of peridotite composition, giving further credence to the idea that the source region of basalt magmas, the Earth's mantle, is dominantly composed of peridotite. Many meteorites are broadly similar to peridotite in composition and provide valuable samples for study. After all, the rocks making up the deep regions of the Earth are not exposed at the Earth's surface and were generally inaccessible for study. One of the earliest accounts of stones falling from the skies is a report from China in 644 B.C., but both the Phoenicians and Egyptians collected meteorites and preserved them in temples as visible signs of material sent to them by their gods. Pliny the Elder recorded the fall of an iron meteorite in Thracia in 476 B.C. "as large as a chariot" and this was said by Anaxagoras to be a fragment of the sun. Aristotle, on the other hand, claimed that meteorites were formed on Earth, but had been carried aloft by a strong wind and subsequently fallen back to Earth. It was on the basis of his idea that such stones were connected with the science of the atmosphere (i.e., meteorology) and hence given their misleading name. Avicenna mentions meteorites from Egypt and Persia, noting that they were stones that fell from heaven to the Earth. One of those he discusses is the sacred black stone, the goal of Islamic pilgrimage, in the Kaaba of Mecca. A number of meteorite falls were recorded during the Middle Ages, but no systematic study was carried out on such stones until 1768 when a commission was set up in France to study a meteorite that had fallen at Lucé in Maine province. The commission, however, concluded that the stone could not have fallen from the heavens. By 1794 the physicist Ernst Chladini had pointed out that meteorites had a chemical composition similar to that of some deep-seated terrestrial rocks in his *On the Origin of the Mass of Iron Found by Pallas in Siberia*. He concluded that meteorites are fragments of cosmic or planetary bodies that are trapped by the Earth's gravitational field, heated by friction with the atmosphere, melt superficially, and then fall to the ground. In 1798 the Italian physicist Ambrogio Soldani showed that meteorites fallen near Siena were indeed different from rocks exposed in the crust of the Earth. Many prominent scholars continued, however, to regard meteorites of telluric or terrestrial origin. Laplace proposed that they were volcanic material ejected from the Moon, whereas others thought they came from terrestrial volcanoes.

Volcanism on the Moon did not seem such a bad idea at the time, considering its cratered and pockmarked visage. Visits of the astronauts to the Moon have shown that there are indeed volcanic rocks there, but they are extremely ancient and not associated with the large craters. Leonardo da Vinci had a true explorer's vision in perceiving the broad similarities between the Earth and the Moon: "If you were where the moon is, it would appear to you that the sun was reflected over as much of the sea as it illuminates in its daily course, and the land would appear amid this water like the dark spots that are upon the moon, which when looked at from the Earth presents to mankind the same appearance that our Earth would present to men dwelling in the moon."[10] Galileo then discovered through his telescope that the Moon "is not robed in a smooth and polished surface, but is rough and uneven, covered everywhere, just like the Earth's surface, with huge prominences, deep valleys and chasms." In 1665 Robert Hooke had first proposed the presence of volcanoes on the Moon.[11] By dropping objects into a "well temper'd mixture of Tobacco-pipe clay and Water" and observing "a pot of boyling Alabaster," he attempted to determine if the craters were of volcanic origin or the result of meteorite impacts. He had observed the lunar craters telescopically (Figure 16.2), and although he was able to reproduce their form during his impact experiments in the laboratory, he considered it very unlikely that they could be produced by the impact of meteorites, "for it would be difficult to imagine whence these bodies should come." Hooke therefore concluded that the lunar craters were of volcanic origin.

The general appearance of the craters on the Moon, as seen by observers in the eighteenth century, also led to the idea that they were of volcanic origin. Georg Christoph Lichtenberg of Göttingen pointed out the similarity of Vesuvius and the mountains on the Moon in 1781; similarly the German scholar U. T. Aepinus described the lunar surface as seen with

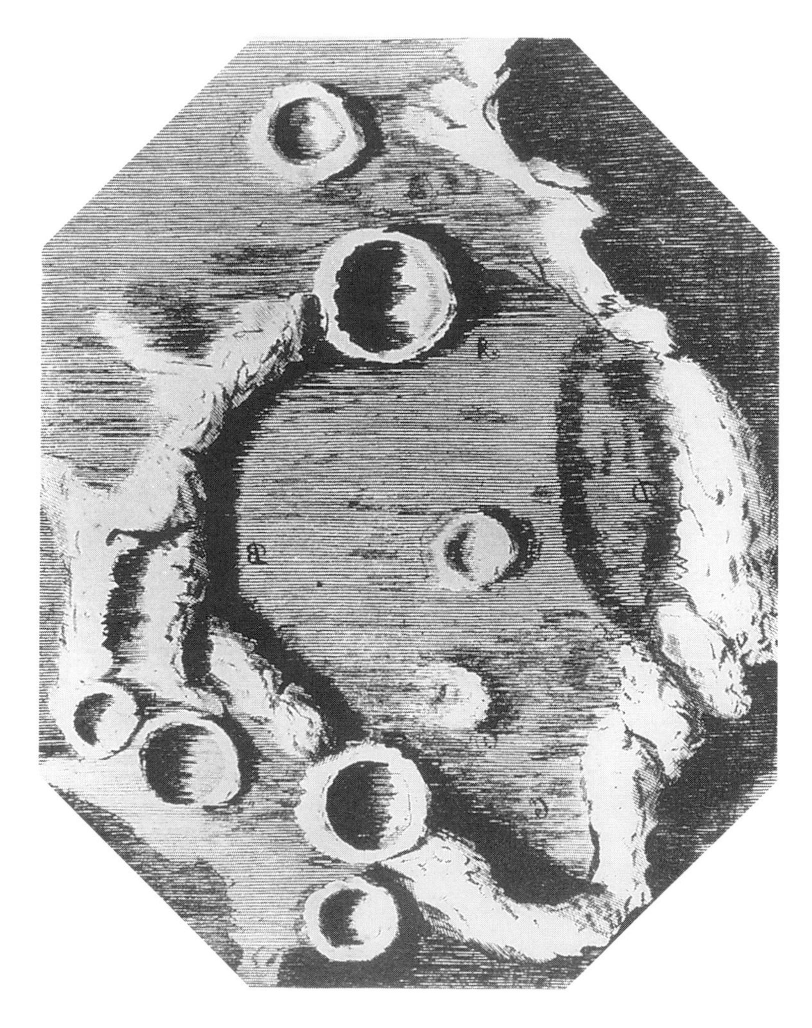

16.2. The British chemist Robert Hooke made this drawing of craters on the Moon (1665). He reproduced their form by dropping bullets into a basin filled with plaster of alabaster, but nevertheless concluded that they were more likely craters of volcanic origin because he found it unlikely that there were meteorites of sufficient size and abundance in the heavens to produce such impacts.

a telescope, arguing that the mountains were volcanic craters. The volcanic hypothesis was further fueled by William Herschel in 1783 who claimed to have observed active volcanoes on the moon, reporting the sighting of glows or two points of light. In 1785, however, Immanuel Kant pointed out that they were much larger than any volcanic features known on Earth; the lunar crater Tycho is, for example, about 225 km (140 mi) in diameter.[12] Kant proved correct in his assessment that the lunar craters are not of volcanic origin; we now know that virtually all are the products of meteorite bombardment.

When the Apollo 12 seismometers detected the first moonquakes in November 1969, scientists got a direct confirmation that the Moon is "dead" inside, harboring no volcanic energy. Moonquakes, it was found, originate about 600 to 800 km (375 to 500 mi) below the surface, are highly localized, and occur at intervals of about fourteen days. Apparently they are triggered by the tidal forces exerted by the Earth. Instead of the typical signature of earthquakes, with a sharp onset and a rapid fall-off, the moonquake vibrations build up to a peak and then gradually fade away over a couple of hours. "The moon rang like a bell" proclaimed one of the lunar scientists. These and other studies demonstrated that the Moon is internally "quiet" and devoid of the creaking, shaking, and groaning of the Earth. The Apollo studies also demonstrated that the Moon has a concentric internal structure, with a crust, mantle, and core. However, it lacks an active magnetic field and there is little or no evidence that it has a molten core, unlike the Earth. Exploration of the lunar surface revealed vast areas of volcanic rocks, known as the Maria, but they turned out to be extremely old, mostly of the order 3.7 billion years. No young volcanic rocks are known on the Moon, nor has volcanic activity ever been witnessed in historical times.[13]

Water in Magmas

The great clouds of gases and steam emitted by volcanoes were long considered to be derived from groundwater, surface waters, such as nearby lakes or streams, or seawater. This was evident, according to many scholars, from the location of volcanoes, which are generally near the ocean or on islands. Thus, the role of water was thought to be crucial in generating explosive eruptions and to have an important effect on the viscosity of magmas. In 1794 Spallanzani recognized that several gases are important in lavas and volcanic regions, including "hydrogenous gas, sulfurated hydrogenous gas, carbonic acid gas, sulfurous acid gas, azotic gas." But another, more powerful agent, he noted, was "water, principally that of the sea," which communicates by passages with the roots of volcanoes. On reaching the subterranean fires it suddenly turns to vapor and the elastic gas expands rapidly, causing volcanic explosions. In support of this he cited examples of the sea receding at the time of terrible eruptions—witness the disturbance of the sea level around the Bay of Naples documented by Pliny the Younger during the eruption of Vesuvius in A.D. 79. Supporting his hypothesis, on a more practical level, Spallanzani cited accidents in glass-making factories, in which molten glass was poured into molds not completely dry or free of water, causing dreadful steam explosions.

Spallanzani worked diligently in trying to establish the source of heat in lava flows. He found their temperature to be higher than that of common fire and comparable to that of a glass-making furnace. He also carried out experiments to determine if sulfur could fuse rocks and turn them into lavas, as had been proposed by the French naturalist Dolomieu. He found, on the contrary, that sulfur did not accelerate the fusion of rocks: "I have thought that sulphurs and petrols give birth to and maintain volcanos, as has been the opinion of the generality of naturalists. It must, however, be confessed, nor do I fear to be contradicted, that such an opinion is very hypothetical, since we are ignorant of what really is the aliment of subterranean fires. It is certain that this aliment, whatever it may be, is, when it burns, in very different circumstances from those of our furnaces, which cannot burn without the presence of atmospheric air, of which the subterranean abysses, where volcanic conflagrations begin, are destitute."[14] He goes on to cite examples of volcanic eruptions in the ocean, where it is clear that air cannot have access to the magma, but hot lavas nonetheless issue from the vents, and speculates that perhaps oxygen gas is present, producing on combustion the heat necessary to melt rocks: "It is not improbable that water united with fire may produce combinations impossible to be attained by human art."

In 1811 the Italian scholar Scipio Breislak also described volcanic rocks and other associated phenomena in Italy.[15] He was a Plutonist, but not one of the dry magma school, and attributed the fluidity of magmas to the joint action of heat and water. The importance of water was also supported by chemical analyses; by 1824 Knox had established by experiment that all volcanic rocks contain some water bound up in the minerals or the rock.

In *Considerations on Volcanos* (1825), Scrope attributed the fluidity of magmas to water: "There can be little doubt that the main agent ... consists in the expansive force of elastic fluids struggling to effect their escape from the interior of a subterranean mass of *lava*, or earths in a state of liquification at an intense heat." As described previously, these "elastic fluids" Scrope considered to be mainly steam and other volcanic gases. The expansion of gas in the magma led to its rise in the Earth's crust and to violent and explosive eruption on reaching the surface.[16]

Scrope discussed in some detail the evolution of steam in magmas at depth and was perhaps the first to point out that under great pressure, water would be dissolved in the melt, but on decrease in pressure or increase in temperature the water would be vaporized, leading to explosive eruption. From the increase of temperature with depth in mines, Scrope concluded that at great depth the Earth was at an intense heat, and that this great accumulation of caloric in the deep Earth led to continued flow of caloric toward the surface, in order for the heat to attempt to attain equilibrium. Thus, Scrope considered the formation of magma as due to the passage of caloric by conduction from depth to upper levels in the Earth, where the caloric led to melting of rocks. Scrope also proposed that the fluidity of molten rocks resulted from water: "There is every reason to believe this fluid to be no other than the vapour of water, intimately combined with the mineral constituents of the lava, and volatilized by the intense temperature to which it is exposed when circumstances occur which permit its expansion." The water of the oceans, he theorized, was derived from the interior of the Earth due to degassing through volcanoes. Large quantities of water "remain entangled interstitially in the condensed matter" in the deep-seated rocks, and the escape of steam during volcanic eruptions carries off an immense amount of caloric, leading to rapid cooling and consolidation of magmas at the surface.[17] Later Charles Lyell adopted Scrope's view on the importance of water in magmas.

By 1865 the French geologist M. Fourque had measured the amount of water in the volcanic products of Etna and estimated the amount of steam discharged from the vent. In the latter part of the nineteenth century geologists began to develop more sophisticated models, proposing that the steam emitted by volcanic eruptions was primordial or of deep-seated origin, rather than just recycled surface waters (Figure 16.3). Among them was the Reverend Osmond Fisher, who put forward this hypothesis in *Physics of the Earth's Crust* (1881): Volcanic gases were indeed original constituents of magma.[18] The potential role of escape of water from volcanoes in the formation of the oceans began to surface as an important hypothesis. The German geologist Eduard Suess (1831–1914) wrote in *Face of the Earth* that all water in the oceans and atmosphere came from the outgassing of the interior of the Earth. The concept of a volcanic recycling of water from the ocean to the atmosphere and back again to the deep Earth was proposed by the British geologist John Judd (1881), who pointed out that volcanoes are generally located near the ocean and speculated that fissures may

16.3. This photograph of Vesuvius in April 1872 may be the earliest taken of a volcanic eruption. The great steam emission during such volcanic eruptions impressed workers like Fisher and Tschermak, who maintained that the steam was derived from primordial water, originating deep within the planet (author's collection).

transport seawater from the ocean to the magma at depth. "Volcanic outbursts" could be due to water finding its way down to a highly heated rock mass, lowering the melting temperature and causing melting.[19] Judd considered high temperature and the presence of water and gas the key factors of volcanic action and argued that magmas can absorb or dissolve large quantities of water, which escapes violently during explosive volcanic eruption, as Spallanzani had first pointed out. The magmas could absorb water or gases either initially, as primordial gases during formation of the globe, or at any stage in geologic history, due to infiltration of water into the Earth's crust.

With the widespread recognition of water in magmas, it was logical to consider its role in explosive eruptions. In the Age of Steam during the Industrial Revolution of the nineteenth century, John Judd pointed out that "a volcano is a kind of great natural steam-engine." Bonney (1899) also compared volcanoes to a boiler, emphasizing that the steam in magma is the main explosive force in an eruption. He noted that the volume of steam is nearly 1700 times that occupied in the form of water, an enormous expansive force, adequate to account for volcanic explosions (Figure 16.4). The origin of volcanic water, according to Bonney, is related to the proximity of volcanoes to the ocean and the percolation of rainwater into the magma.[20] He stressed the importance of the addition of water to lower the melting point of rocks and followed Fisher and others in attributing eruption to the presence of water. When water is depleted or withdrawn from the system, the eruption ceases.

In the latter part of the eighteenth century, Dolomieu had determined that many deposits in Italy and on Sicily, which at first sight looked like normal stratified sediments, were in fact consolidated volcanic ash deposits, the products of fallout from explosive eruptions. The disruption of magma as it is blown into the air leads to fragmentation and the formation of volcanic ash, tephra, scoria, and other forms of pyroclastic material. We now know

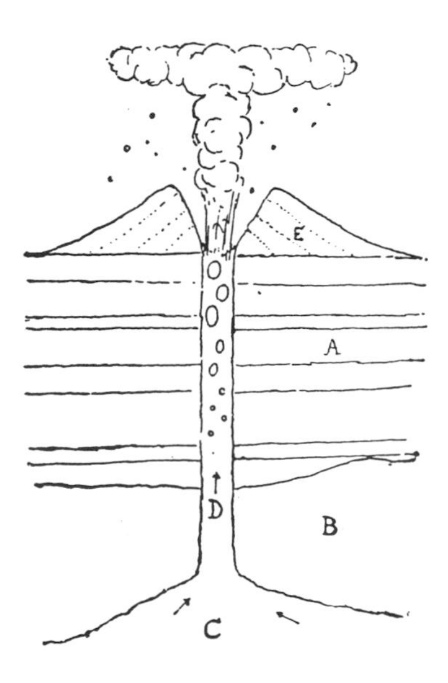

16.4. Bonney (1899) and others realized that water dissolved in the magma at great depth would form steam or gas bubbles during the rise of the magma up through the Earth's crust. These bubbles would grow in size as pressure decreased until they burst explosively near the surface and disintegrated the magma into volcanic ash particles.

that the primary agent of this disruption is the explosive expansion of steam. Sartorius von Waltershausen a true pioneer in the study of pyroclastic rocks, was the first to attribute their formation to the effects of water on the magma (1853).[21] He proposed that magma rises and erupts due to the pressure of water vapor escaping from the magma, and that the volume increase of water vapor was also responsible for the fragmenting process—the formation and ejection of pyroclasts such as pumice and ash. He also recognized that peculiar volcanic rock deposits can result when magma enters the ocean or is erupted under water. During travels in Sicily in 1835 he first came across a brown tuff, a rock with a homogenous mass, composed largely of a single mineral. He named the rock "palagonite" after the nearby town of Palagonia, and by chemical analysis determined it as one unusually rich in water and iron, with about 12 to 23% wt water. He became curious about the rock's origin and, noting that it was often associated with marine deposits, proposed that it was a volcanic product that forms thick layers in many submarine volcanic formations. In the summer of 1846 he and Robert Bunsen studied a number of large palagonite mountains in Iceland and noted that a zone of thick palagonite tuffs stretches across Iceland from the southwest to the northeast. Often in association with palagonite, was a black, water-free, glass-like material, similar to obsidian, which he gave the name *sideromelan*. He was able to show that palagonite is basically sideromelane or basaltic glass, which has taken up water and, from its geologic setting, palagonite as a product of shallow subaquatic or submarine basaltic eruptions. Von Waltershausen was correct in this deduction; we now know that the Icelandic palagonite rocks are formed by volcanic eruptions below the thick ice cap that covered Iceland during the last ice age and that the Italian palagonites were erupted in the ocean.

Bunsen, on the other hand, disagreed with von Waltershausen, considering the palagonite tuffs the products of basaltic rocks metamorphosed in the presence of much water and carbonate.[22] He showed by chemical analysis that the tuffs were virtually identical to basalt lava after the high water content had been subtracted. His study was not restricted to Icelandic rocks because Charles Darwin had given him samples of palagonite from the Cape Verde Islands.

Another volcanic rock of basaltic composition is lava-like but composed of rounded or pillow-like forms. The origin of this rock does not have a bearing on water in magmas, but rather on magma in the water. The identification of *pillow lava* dates back at least to the 1870s; later it was proposed that it forms due to submarine eruption, on the basis of observations in Italy.[23] The British geologist Tempest Anderson observed an eruption in Samoa, noting that when the basaltic lava was flowing into the sea the submarine component of the lava formed bulbous masses and lobes with the shape of pillows.[24] By 1914 an aqueous origin for pillow lavas had been established. They were already known to be present in the geologic record in association with marine sediments in Scotland and also as products of subglacial eruptions, such as in Iceland. Oceanographers discovered only in the 1960s that basaltic pillow lava forms the floor of most of the world's oceans and is thus the Earth's most abundant volcanic rock but also the most remote for study.[25]

When high-pressure experiments of melting were first carried out at the beginning of the twentieth century, it became evident that magmas residing deep in the Earth can contain much more water in solution and that this water must be liberated when they are erupted at the surface. In 1903 C. Doelter was one of the first to propose that magmas at great pressure in the Earth's crust may dissolve water and that the magmas may become explosive when they reach the surface of the Earth. In support of this theory, the French geologist Armand Gautier (1900–1906) carried out laboratory experiments on volcanic activity, suggesting that magma rises in the crust as a result of gas expansion, and attributed the violence of volcanic eruptions to the explosive liberation of water from the magmas. By the early twentieth century the fundamental ideas about the causes of explosive volcanism and the importance of water had thus been established.[26]

Plate Tectonics and Magma Generation

If you sail southwest from the Reykjanes peninsula of Iceland, you follow the crest of a great volcanic mountain ridge, the Reykjanes Ridge, and the scene of many a submarine volcanic eruption. To the south the ridge merges with the Mid-Atlantic Ridge, one of the major branches of the world-girdling mid-ocean ridge system. In 1963 a great discovery was made on the Reykjanes Ridge, one that led to the transformation of geology from a purely descriptive science to one based on a fundamental understanding of how the Earth works. Before 1963 geologists were like timekeepers of the Earth who could listen to the tick of the clock and watch the hands move over the dial, but in that year, some of the actual works of the clock were revealed, marking the birth of the theory of seafloor spreading and plate tectonics. One of the beauties of this revolution was that it gave proof of the mechanism that Arthur Holmes had proposed in 1928 by which the Earth's mantle convects, undergoes decompression, and melts. It provided the solution to the riddle of melting in the Earth.

The concept of plate tectonics has had a long gestation period. It began with the realization in the seventeenth century that volcanoes and earthquakes are distributed in great linear belts over the Earth. The systematic trends of volcanoes and earthquake zones, and the arrangement of the continents, eventually led to the theory known as continental drift in the early twentieth century. The earliest map showing the global distribution of volcanoes was presented by Athanasius Kircher in the *Mundus Subterraneus* (1665), an invaluable record in which he identified the volcanic regions of Italy, Iceland, the Andes of South America, the East Indies (Indonesia), Japan, and Kamchatka, among others (Figure 16.5). With the spread of geographic knowledge, a more refined picture gradually emerged, one of the systematic, linear, or curvilinear arrangement of volcanoes on the surface of the globe.

Ideas on the relationship between volcanic activity and major structures and tectonics date to the great German geographer Alexander von Humboldt (1822), who pointed out that the linear arrangement of volcanoes on the Earth was a proof that the mechanism that generates volcanism was not superficial, but a deep-seated feature. As more information accumulated on the global distribution of earthquakes and volcanic eruptions, Robert Mallet

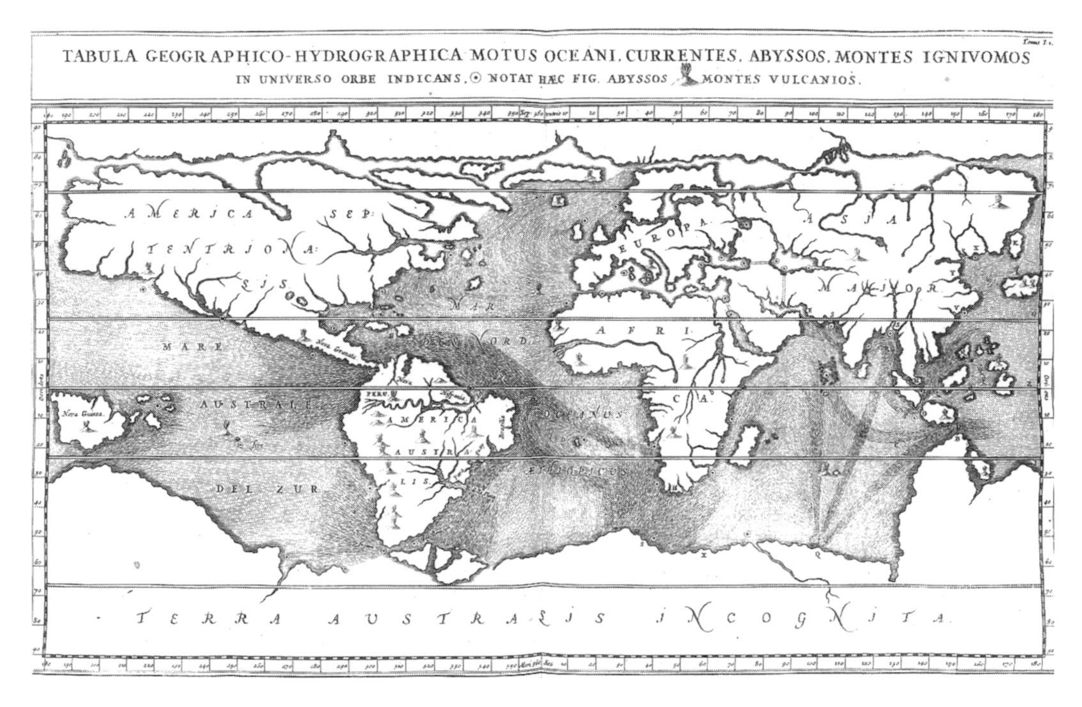

16.5. The first map showing the distribution of volcanoes on the Earth was drawn by the German Jesuit priest Athanasius Kircher in 1665 and published in his important work *Mundus Subterraneus*.

(1858) compiled the first map showing both of these features in a "vast loop or band round the Pacific," indicating a strong coincidence in the geographic distribution. [27] He also recognized that most of the Earth, especially the interior of the continents, is seemingly free of earthquakes and volcanic activity. He divided the Earth into a series of basins, separated by "girdling ridges" that are present even on the ocean floor: "It is along these girdling ridges, whether mountain-ranges or mere continuous swelling elevations of the solid, which divide these basins beneath the ocean surface one from the other, that all the volcanoes known to exist upon the earth's surface are found, dotted along these ridges or crests in an unequal and uncertain manner." He regarded these ridges as "sub-oceanic linear volcanic ranges" that mark "the great lines of fracture of the earth's crust." Mallet's recognition of the distribution of these geological structures, which are now basic to the theory of plate tectonics, was probably the earliest.

This recognition of the linearity of volcanoes and earthquake zones coincided with a mobilistic view of the Earth toward the end of the century, a view based both on geophysical reasoning and geologic evidence (Figure 16.6).[28] In his book *Physics of the Earth's Crust* (1881), the Reverend Osmond Fisher had argued that cooling of the Earth could not be brought about solely by the process of conduction, as had been proposed by Kelvin, but was in part rendered by convection currents in a plastic substratum. In Fisher's view, these convection currents led to lateral movement of the overlying crust, accounting for much of the

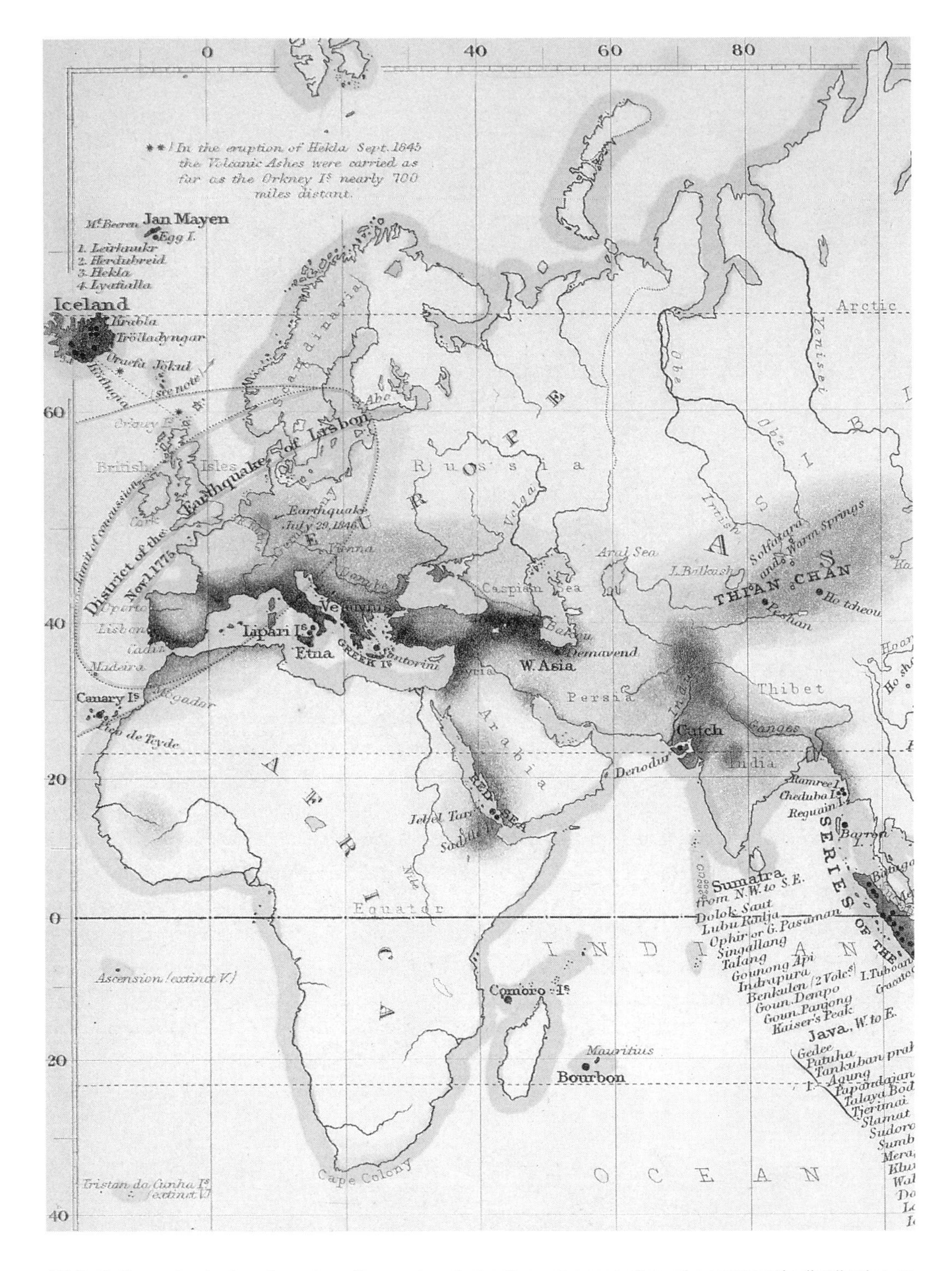

16.6. By the early nineteenth century, it was clear that active volcanoes showed a systematic distribution on the Earth's surface, forming lines or arcs, especially around the Pacific Ocean. Detail of 1850 engraving by Augustus Petermann, depicting the location and names of active volcanoes and the regions of earthquake activity (shaded).

character of the Earth's surface, including linear arrangement of volcanoes, earthquakes, and mountain chains.

The ideas of a mobile Earth were first crystallized into a coherent theory of continental drift in 1912 by the German meteorologist Alfred Wegener (1880–1930). Wegener had been struck by the similarity of the coastlines of the continents surrounding the Atlantic Ocean, particularly the perfect jigsaw fit between the coasts of Africa and South America.[29] When he published his theory in *Die Entstehung der Kontinente und Ozeane* (The Origin of Continents and Oceans, 1915), Wegener marshaled a number of lines of evidence to support his case for a mobile Earth, such as the great similarities of rocks and matching paleobotanical provinces on either sides of the Atlantic. Reaction to this revolutionary theory was generally hostile in Britain and North America and remained so until the 1960s, when the discovery of seafloor spreading on the Reykjanes Ridge proved him right.

One of those siding with Wegener was Arthur Holmes, who greatly strengthened the theory by proposing in 1928 a more plausible thermal convection mechanism for crustal movement than Wegener had put forward. Holmes's mechanism involved the flow or upwelling of the Earth's mantle, providing a process whereby decompression of the hot mantle would occur, leading to melting and volcanism.[30] As previously mentioned, Arthur Holmes began his career in physics at Imperial College in London, where he worked with Lord Rayleigh, pioneering the uranium-lead method for the radioactive dating of rocks.[31] Realizing the great potential of the recent discovery of both radioactive dating and radiogenic heat sources for understanding the Earth, he abandoned physics and turned his great talents to geology. He was quick to see the significance of Wegener's continental drift theory in relation to his own work on heat generation within the Earth by radioactive decay and became convinced that this heat output was much larger than could be explained by pure conduction. Instead, he proposed that the excess heat generated by the radioactive natural reactor was discharged by thermal convection in the Earth's mantle (Figure 16.1). Continental drift was an inevitable consequence of this circulation, but the process was much wider in scope: it involved the lateral motion of the oceanic crust as well. By 1931 he had constructed a complete theory of continental drift, seafloor spreading, and subduction based on this mechanism.

Holmes differentiated between a largely granitic crust in the continents and a basaltic crust in the oceans. He considered the mantle beneath the oceanic crust as most likely of peridotite composition and argued that, although it was probably crystalline and thus highly rigid, it had some of the properties of a fluid in the context of its large-scale dimensions. He proposed that the convection cells have dimensions at the scale of the mantle, some 2900 km (1813 mi) in depth, and that their ascending limbs rise beneath the oceans where they generate oceanic "swells" analogous to our modern concept of mid-oceanic ridges. These he envisioned as "stretched regions or disruptive basins" in which the formation of new basaltic ocean floor occurs, involving the discharge of a great deal of heat. He proposed that ascending currents were accompanied by decompression and partial melting of peridotite to form basaltic magma, giving rise to volcanism. Descending currents were caused in the mantle by the sinking of high-density rocks. The process he outlined was essentially the same as the modern concept of subduction: the convergence of crustal plates on the Earth leads to underthrusting of one plate and its descent deep into the mantle. Holmes estimated that the velocities of the convection current were on the order of 5 cm per year, which, as discovered in the 1960s, is perfectly within the range of typical seafloor spreading rates. He proposed that the oceanic basin now occupied by the Atlantic had taken some 100 million years to form by the process of continental drift and convective flow in the mantle. Again he was very close; we now know that the actual figure is about 160 million years. Although widely accepted in Great Britain, Holmes' ideas on mantle convection and a mobile Earth were generally not embraced elsewhere for some time. When the idea of seafloor spreading was finally accepted by the majority of Earth scientists in the 1960s Holmes's fundamental contribution was frequently ignored. One is reminded of the chilling words of Sir William Osler: "In science the credit goes to the man who convinces the world, not to the man to whom the idea first came." Is it more important in science to convince than to discover?

Holmes continued to develop his theory of mantle convection and crustal drift in his great treatise *Principles of Physical Geology*, first published in 1944. This book contained a final chapter devoted to continental drift and mantle convection. The 1960 edition was my first textbook when I began my study of geology in 1962. My professors treated this final chapter as an afterthought, but I was intrigued because it seemed to explain the rifting features of my home country, Iceland, sitting astride the Mid-Atlantic Ridge. The volcanic crust of Iceland is split asunder by a rift valley, with many gaping fissures. Soon after I returned to resume my classes in the fall of 1963, a three-page paper appeared in the journal *Nature* by the British geologists Fred Vine and Drummond Matthews that provided convincing proof of Arthur Holmes's theory. This proof was made possible by the measurement of magnetic properties of the Earth's crust beginning in 1961.[32] As these measurements began to accumulate, a regular pattern emerged, showing a symmetrical distribution of alternately normally and reversely magnetized stripes of oceanic crust flanking the mid-ocean ridges of the world. Vine and Matthews came up with the brilliant idea that the creation of new ocean floor by volcanism at the mid-ocean ridge acts as a magnetic tape recorder of the episodic reversals in the Earth's magnetic field. This magnetic record was so incredibly regular throughout the ocean floor that it was overwhelming evidence of seafloor spreading and continental drift. The discovery of plate tectonics has now provided the long-sought mechanism for decompression in the Earth's mantle—the most important process that brings about melting and generation of magmas. It is clearly the mechanism that generates magmas below the mid-ocean ridges of the Earth—and is thus responsible for most of our planet's volcanism.

But volcanism also occurs in areas where plates collide—the subduction zones—caused by the underthrusting of one plate by another. Above the subduction zones reside most of the Earth's best-known volcanoes, including Mount St Helens, Vesuvius, Krakatau, Fuji, and Mount Pelée. The paradox of subduction zones is that they are regions where a cold crustal plate is thrust at an angle down into the Earth's mantle—why should this result in melting and volcanism above? We are still not sure about the answer, but the melting of the mantle below the subduction zone volcanoes is probably caused by two factors. First, the descending plate brings water into the Earth's mantle, which lowers its melting point and facilitates magma generation. Second, the descent of the cold plate stirs up the mantle and causes hotter mantle material to rise towards the surface, resulting in a decrease of pressure and setting in motion a melting process identical to that occurring below the mid-ocean ridges of the world.

We think we have discovered the secrets of volcanism on the Earth, but other worlds bring new surprises. During a 1979 flyby of Io, one of the satellites of Jupiter, by the *Voyager* spacecraft, the scientific community on Earth was astonished to observe enormous volcanic eruptions emanating from its surface. We had become accustomed to the idea that active volcanism was an exclusive property of the Earth; all space misssions to date had shown that Mars, Venus, and the Moon appeared to be volcanically "dead." Io, on the other hand, exhibited volcanism on a grand scale, with over 600 volcanoes recognized, spaced about 250 km apart on the average, and some of them spouting eruption plumes up to 400 km height, or ejecting their products ten times higher than any known eruption on Earth. The distribution of volcanism on Io is best compared to the random "boils" that appear on the surface of a bubbling hot soup in a pot. The *Galileo* spacecraft in 1996 confirmed these findings of exceptional volcanism and the results also showed that melting on Io is probably of a very different cause from melting in the Earth. Perhaps the most astonishing result of Io exploration is the extremely high temperature of some of its magmas, up to 1500°C, or 300°C hotter than the hottest terrestrial magmas.[33] In addition, the *Galileo* spacecraft showed that Io may have a magnetic field—a good indication of a hot liquid core. The heat flow from Io is about one hundred times higher than what can be accounted for by radioactive decay and it occurs from a planet that appears devoid of plate tectonics. Where is all this heat coming from? Since it is not related to the mantle upwelling associated with the spreading of crust and formation of mid-ocean ridges, some other fundamental process must be at work—and it turned out to be the tide. Even before the discovery of extensive volcanism on Io, the astronomer Stanton Peale had predicted that the tidal pull from the giant planet Jupiter and its nearby moon Europa was so immense that it led to the heating of Io's interior.[34] The pull of Jupiter—a planet with almost three hundred times the mass of the Earth—continually deforms and flexes the mantle of Io, creating a tidal bulge of solid rock that sweeps over the surface of the planet. Daily, Io is squeezed hard by the tidal pull, and magma squirts out like juice from an orange. The vast energy of this tidal force is dissipated as frictional heating within Io, leading to melting of its deep layers. Thus, now that we are at last understanding the melting processes in our Earth, we are suddenly faced with an entirely different melting process on other planets. What other mysteries await us as we venture further out into the solar system?

Notes

Preface

1. Sigurdsson, H., S. Cashdollar, and R. S. J. Sparks, 1982. "The Eruption of Vesuvius in A.D. 79: Reconstruction from Historical and Volcanological Evidence." *Amer J. Archeology* 86: 39–51. Sigurdsson, H., S. Carey, W. Cornell, and T. Pescatore. 1985. "The Eruption of Vesuvius in A.D. 79." *National Geographic Research* 1: 332–87.
2. Laudan, R. 1987. *From Mineralogy to Geology. The Foundations of a Science 1650–1830*. Chicago: University of Chicago Press.
3. Krafft, M. 1991. *Les Feux de la terre*. Paris: Gallimard.
4. Brush, S. G. 1979. "Nineteenth-century Debates about the Inside of the Earth: Solid, Liquid or Gas?" *Annals of Science* 36: 225–54.
5. Laudan, *From Mineralogy to Geology*.
6. Kushner, D. S. 1990. "The Emergence of Geophysics in Nineteenth-Century Britain." Ph.D. thesis, Princeton University.

Chapter 1

1. Roberts, J. M. 1976. *History of the World*. New York: A Knopf.
2. Gowlett, J. A. J. 1993. *Ascent to Civilization*. New York: McGraw-Hill.
3. Frazer, J. G. 1930. *Myths of the Origin of Fire*. London: Macmillan.
4. Leakey, M. D. 1979. "Footprints in the Ashes of Time." *National Geographic*, April: 446–57.
5. Johanson, D., and M. Edey. 1981. *Lucy; The Beginnings of Humankind*. New York: Simon and Schuster.
6. Leakey, R., and R. Lawin. 1992. *Origins Reconsidered*. New York: Doubleday.
7. Stamatopoulos, A. C., and P. C. Kotzias. 1990. "Volcanic Ash in Ancient and Modern Construction" in D. A. Hardy (ed.), *Thera and the Aegean World*, vol. III. London: The Thera Foundation, 491–501.

8. Morris, J. B. 1928. *Hernando Cortes: Five Letters 1519–1526*. London: Routledge, 32.
9. Hopkins, W. 1839. "On the Phenomena of Precession and Nutation, Assuming Fluidity of the Interior of the Earth." *Phil. Trans. Roy. Soc. London*: 381–423. Fisher, O. 1881. *Physics of the Earth's Crust*. London: Macmillan.
10. Shelley, M. W. 1831. *Frankenstein, or the Modern Prometheus*. London: Colburn and Bentley.
11. Guiot, J. 1992. "The Climate of Central Canada and Southwestern Europe Reconstructed by Combining Various Types of Proxy data" in C. R. Harington (ed.), *The Year Without Summer? World Climate in 1816*. Ottawa: Canadian Museum of Nature, 291–308.
12. Manley, G. 1946. "Temperature Trend in Lancashire 1753–1945," *Quart. J. Roy. Met. Soc.* 72: 1–21.
13. Ladurie, E. L. R. 1971. *Times of Feast, Times of Famine. A History of Climate Since the Year 1000*. New York: Farrar, Straus.
14. Neumann, J. 1992. "The 1810s in the Baltic Region, 1816 in Particular: Air Temperatures, Grain Supply and Mortality" in Harington, *The Year Without Summer?*, 392–415.
15. Baron, W. A. 1992. "1816 in Perspective: The View from the Northeastern United States" in Harington, *The Year Without Summer?*, 124–44.
16. Post, J. D. 1977. *The Last Great Subsistence Crisis in the Western World*. Baltimore: Johns Hopkins Univ. Press.
17. Forsyth, P. Y. 1988. "In the Wake of Etna, 44 BC." *Classical Antiquity* 7: 49–57. Univ. of California Press.
18. Metrich, N., and R. Clocchiatti. 1989. "Melt Inclusion Investigation of the Volatile Behavior in Historic Alkali Basaltic Magmas of Etna." *Bull. Volcanol.* 51: 185–98.
19. Sigurdsson, H. 1982. "Volcanic Pollution and Climate: The 1783 Laki Eruption." *EOS* 63: 601–02.
20. Franklin, B. 1784. "Meteorological Imaginations and Conjectures." The Literary and Philosophical Society of Manchester. *Memoirs*, vol. 2: 373–77.

Chapter 2

1. Mellaart, J. 1965. *The Earliest Civilizations of the Near East*. New York: McGraw-Hill. Mellaart, J. 1967. *Çatal Hüyük: A Neolithic Town in Anatolia*. New York: McGraw-Hill.
2. Mellart, 1965. *Earliest Civilizations*.
3. Poole, L., and G. Poole. 1962. *Volcanoes in Action: Science and Legend*. New York: McGraw-Hill.
4. Vitaliano, D. B. 1973. *Legends of the Earth*. Bloomington: Indiana Univ. Press.
5. Stephens, J. L. 1969. *Incidents of Travel in Central America, Chiapas and Yucatan*, vol. 2. New York: Dover.
6. Secor, R. J. 1981. *Mexico's Volcanoes, A Climbing Guide*. Seattle: The Mountaineers.
7. Coleman, S. N. 1946. *Volcanoes New and Old*. New York: The John Day Company.
8. Greene, M. T. 1992. *Natural Knowledge in Preclassical Antiquity*. Baltimore: Johns Hopkins Univ. Press.
9. Graves, R. 1982. *The Greek Myths*, Vol. 1. Harmondsworth, England: Penguin.
10. Turner, A. K. 1993. *The History of Hell*. New York: Harcourt Brace.
11. Johnson, R. W., and N. A. Threlfall. 1985. *Volcano Town. The 1937–43 Rabaul Eruptions*. Robert Brown & Associates, Australia.
12. Westervelt, W. D. 1963. *Hawaiian Legends of Volcanoes*. Rutland, Vt.: Charles E. Tuttle.
13. Ibid.

Chapter 3

1. Sigurdsson, H., S. Carey, and J. D. Devine. 1990. "Assessment of Mass, Dynamics and Environmental Effects of the Minoan Eruption of Santorini Volcano" in D. A. Hardy (ed.), *Thera and the Aegean*

World, vol. III. London: The Thera Foundation, 100–12.

2. Guichard, F., S. Carey, M. A. Arthur, H. Sigurdsson, and M. Arnold. 1993. "Tephra from the Minoan Eruption of Santorini in Sediments of the Black Sea." *Nature* 363: 610–12.
3. C. G. Doumas. 1983. *Thera, Pompeii of the Ancient Aegean*. London: Thames and Hudson. For an up-to-date and comprehensive view of research related to the Thera eruption see the three-volume Proceedings of the Third International Congress, Santorini: D. A. Hardy (ed.), *Thera and the Aegean World*. London: The Thera Foundation, 1990.
4. Sigurdsson, H., S. Carey, and C. Mandeville. 1991. "Krakatau: Submarine Pyroclastic Flows of the 1883 Eruption of Krakatau Volcano." *National Geographic Research* 7: 310–27. Sigurdsson, H., S. Carey, C. Mandeville, and S. Bronto. 1991. "Pyroclastic Flows of the 1883 Krakatau Eruption." *EOS*: 377–81.
5. Luce, J. V. 1969. *Lost Atlantis*. New York: McGraw-Hill.
6. Niemeier, W. D. 1980. "Die Katastrophe von Thera und die spätminoische Chronologie." Jahrbuch des Deutschen Archäologischen Instituts 95: 1–76. Doumas, C. G, 1989. "Archaeological Observations at Akrotiri Relating to the Volcanic Destruction" in Hardy, *Thera and the Aegean World*, 48–50.
7. Limbrey, S. 1990. "Soil Studies at Akrotiri" in Hardy, *Thera and the Aegean World*, 377–83.
8. Manning, S. W. 1990. "The Thera Eruption: The Third Congress and the Problem of the Date." *Archaeometry* 32 (1): 91–100.
9. Housley, R. A., R. E. M. Hedges, I. A. Law, and C. R. Bronk, 1980. "Radiocarbon Dating by AMS of the destruction of Akrotiri" in Hardy, *Thera and the Aegean World*, 207–15. Manning, S. W., and B. Weninger. 1992. "A Light in the Dark: Archaeological Wiggle Matching and the Absolute Chronology of the Close of the Aegena Late Bronze Age." *Antiquity* 66: 636–63.
10. Watkins, N. D., R. S. J. Sparks, H. Sigurdsson, T. C. Huang, A. Federman, S. Carey, and D. Ninkovich. 1978. "Volume and Extent of the Minoan Tephra Layer from Santorini Volcano: New Evidence from Deep-sea Sediment Cores." *Nature* 271: 122–26.
11. Marketou, T. 1990. "Santorini Tephra from Rhodes and Kos" in Hardy, *Thera and the Aegean World*, 96–113.
12. Warren, P. M. 1991. "The Minoan Civilization of Crete and the Volcano of Thera." *J. Ancient Chronol. Forum* 4: 29–40.
13. Soles, J. S., and C. Davaras. 1990. "Theran Ash in Minoan Crete: New Excavations on Mochlos" in Hardy, *Thera and the Aegean World*, 89–95.
14. Marinatos, N. 1984. *Art and Religion in Thera*. Athens: D. & I. Mathioulakis.
15. Bennett, J. G. 1963. "Geophysics and Human History: New Light on Plato's Atlantis and the Exodus." *Systematics* 1: 127–56.
16. Vitaliano, D. B. 1973. *Legends of the Earth*. Bloomington: Indiana Univ. Press.
17. Frost, K. T. 1909. *The Lost Continent*. London: *The Times*, 19 February 1909. Frost, K. T. 1909. "The Critias and Minoan Crete." *Journal of Hellenic Studies* 33: 189.
18. Luce, J. V. 1969. *Lost Atlantis*. New York: McGraw-Hill. Galanopoulos, A. G., and E. Bacon. 1969. *Atlantis: The Truth Behind the Legend*. New York: Bobbs-Merrill. C. G. Doumas. 1983. *Thera, Pompeii of the Ancient Aegean*. London: Thames and Hudson.
19. Sarton, G. 1993. *Ancient Science Through the Golden Age of Greece*. New York: Dover.
20. Greene, M. T. 1992. *Natural Knowledge in Preclassical Antiquity*. Baltimore: Johns Hopkins Univ. Press.
21. Graves, R. 1982. *The Greek Myths*, Vol. 1. Harmondsworth, England: Penguin.
22. Herm, G. 1975. *The Phoenicians*. New York: William Morrow.
23. Charles-Picard, G. and C. Charles Picard. 1961. *Daily Life in Carthage*. New York: Macmillan.
24. Talbot, P. A. 1926. *The Peoples of Southern Nigeria*. London.

25. Rappaport, R. 1986. "Hooke on Earthquakes: Lectures, Strategy and Audience." *Brit. J. Hist. Sci.* 19: 129–46.
26. Sigurdsson, H., J. D. Devine, F. M. Tchoua, T. S. Presser, M. K. W. Pringle, and W. C. Evans. 1987. "Origin of the Lethal Gas Burst from Lake Monoun, Cameroun." *J. Volcanol. Geotherm. Res.* 31: 1–16. Sigurdsson, H. 1988. "Gas Bursts from Cameroon Crater Lakes: A New Natural Hazard." *Disasters* 12: 131–46. Sigurdsson, H. 1987. "Lethal Gas Bursts from Cameroon Crater Lakes." *EOS* 68: 570–73.

Chapter 4

1. Thompson, S. J. 1988. *A Chronology of Geological Thinking from Antiquity to 1899*. Metuchen, N.J.: Scarecrow Press.
2. The scholars in Plato's Academy were the first to develop geometry, the science of the measurement of the Earth. "Let none but Geometers enter here" was the motto engraved above the door to the Academy. It was through the development of geometry that the early Greeks, from Plato, Eudoxus, and Aristotle, laid the foundations for geodesy, one of the cornerstones of geophysics.
3. Chester, D. K., A. M. Duncan, J. E. Guest, and C. R. J. Kilburn. 1985. *Mount Etna*. Stanford: Stanford Univ. Press.
4. Greene, M. T. 1992. *Natural Knowledge in Preclassical Antiquity*. Baltimore: Johns Hopkins Univ. Press.
5. The harbor at Cumae was the site of the earliest Greek settlement, about 750 B.C., on the mainland of Italy. In the fourth century B.C. Cumae and the Phlegraean Fields finally came under the rule of Rome. Paget, R. F. 1968: "The Ancient Ports of Cumae." *The Journal of Roman Studies* 58: 150–69.
6. Stothers, R. B., and M. R. Rampino. 1983. "Volcanic Eruptions in the Mediterranean Region Before A.D. 630 from Written and Archaeological Sources." *J. Geophys. Res.* 88: 6357–71. Forsyth, P. Y. 1988. "In the Wake of Etna, 44 B.C." *Classical Antiquity* 7: 49–57. Univ. of California Press.
7. Ferris, T. 1988. *Coming of Age in the Milky Way*. New York: William Morrow.
8. Bovie, P. 1974. *Lucretius: On the Nature of Things*. Translation from the Latin text. New York: New American Library.
9. Adams, F. D. 1938. *The Birth and Development of the Geological Sciences*. New York: Dover.
10. Seneca was Rome's leading intellectual figure during the middle of the first century. As Nero's tutor, he had a relationship with the emperor rather like that of Aristotle with Alexander the Great, but with the difference that the would-be Roman philosopher king turned into an insane tyrant. After having been falsely accused of plotting to overthrow Nero in the so-called Pisonian conspiracy, Seneca tragically took his own life in 65 A.D.
11. Geikie, A. 1910. "Notes on Seneca's 'Quaestiones Naturales' " in J. Clarke, *Physical Science in the Time of Nero*. London: Macmillan, 309–43.
12. Anon. 1961. *Aetna* in *Minor Latin Poets: With Introduction and English Translation by J. Wright Duff and Arnold Duff*. London and Cambridge, Mass.: Loeb Classical Library, 349–419. Paisley, P. B., and D. R. Oldroyd. 1979. "Science in the Silver Age: *Aetna*, a Classical Theory of Volcanic Activity." *Centaurus* 23: 1–20.
13. Beagon, M. 1992. *Roman Nature: The Thought of Pliny the Elder*. New York: Oxford Univ. Press.

Chapter 5

1. The Roman economy flourished during the reign of Vespasian and there is much evidence of increased building activity in Italy at this time, including restoration of the temple of Mater Deum in Herculaneum in A.D. 76 that had been severely damaged in the earthquake of A.D. 62. Although his power was purely military, Vespasian reestablished relations with the Senate and renewed the aristoc-

racy. He placed new men at the head of the military, including the scholar Pliny the Elder, who held office as a military tribune and in A.D. 71 became commander of the Roman fleet at Misenum near Vesuvius.

2. Cassius, D. 1927. *Roman History*, Translated from the Latin by Ernest Cary. New York: Macmillan, xvi, 20–24.
3. Grant, M. 1975. *The Twelve Caesars*. New York: Charles Scribner's.
4. Deiss. J. J. 1985. *Herculaneum*. New York: Harper & Row.
5. Healy, J. F. 1986. "Pliny on Mineralogy and Metals" in R. French, and F. Greenaway, *Science in the Early Roman Empire*. N.J.: Barnes and Noble, 111–46.
6. Vogel, J. S., W. Cornell, D. E. Nelson, and J. R. Southon. 1990. "Vesuvius/Avellino, One Possible Source of Seventeenth Century BC Climatic Disturbances." *Nature* 344: 534–37. Manning, S. W., and B. Weninger. 1992. "A Light in the Dark: Archaeological Wiggle Matching and the Absolute Chronology of the Close of the Aegena Late Bronze Age." *Antiquity* 66: 636–63.
7. Stothers, R. B., and M. R. Rampino. 1983. "Volcanic Eruptions in the Mediterranean Region Before A.D. 630 from written and Archeological Sources." *J. Geophys. Res.* 88: 6357–71.
8. Grant, M. 1968. *Gladiators*. New York: Delacorte Press.
9. Ibid.
10. Corti, E. C. C. 1940. *Untergang und Aufstehung von Pompeji und Herculaneum*. Munich: Bruckmann.
11. Misenum had become the headquarters of the western imperial navy under Augustus. Over 10,000 sailors were based here in Nero's time, mostly recruited from the colonies in Egypt, Thrace, and parts of the Greek East; see J. H. D'Arms, 1970. *Romans on the Bay of Naples*, Cambridge: Harvard Univ. Press.
12. Radice, B. (ed.) 1963. *The Letters of the Younger Pliny*. Harmondsworth, England: Penguin.
13. Sigurdsson, H., S. Cashdollar, and R. S. J. Sparks. 1982. "The Eruption of Vesuvius in A.D. 79: Reconstruction from Historical and Volcanological Evidence." *Amer. J. Archeology* 86: 39–51. Sigurdsson, H., S. Carey, W. Cornell, and T. Pescatore. 1985. "The Eruption of Vesuvius in A.D. 79." *National Geographic Research* 1: 332–87.
14. Bigelow, J. 1858. "On the Death of the Elder Pliny." *Memoirs of the American Academy of Arts and Sciences* 6: 223–27. Sirks, M. J., and C. Zirkle. 1964. *The Evolution of Biology*. New York: Ronald Press, 55. Zirkle, C. 1967. "The Death of Gaius Plinius Secundus (23–79 AD)." *Isis* 58: 553–59. Haywood, R. M. 1952. "The Strange Death of the Elder Pliny." *The Classical Weekly* 46: 1–3. Lipscomb, H. C. 1954. "The Strange Death of the Elder Pliny." *The Classical Weekly* 47: 74.
15. Winkes, R. 1982. "Roman Paintings and Mosaics." *Catalogue of the Classical Collection*. Providence: Rhode Island School of Design.
16. Jashemski, W. F. 1979. *The Gardens of Pompeii, Herculaneum and the Villas Destroyed by Vesuvius*. New Rochelle, N.Y.: Caratzas Brothers.
17. Ferrero, F. L. 1930. "That Amphora and the Death of Pliny the Elder." *Art and Archaeology* 29: 50–55.
18. Ibid.
19. Greco, G. 1994. "The Greeks on the Shores of Ancient Campania." International Association of Sedimentologists Meeting, DeFrede-Napoli.
20. Sigurdsson, H., S. Cashdollar, and R. S. J. Sparks. 1982. "The Eruption of Vesuvius in A.D. 79: Reconstruction from Historical and Volcanological Evidence." *Amer. J. Archeology*. 86: 39–51. Sigurdsson, H., S. Carey, W. Cornell, and T. Pescatore. 1985. "The Eruption of Vesuvius in A.D. 79." *National Geographic Research* 1: 332–87. Sigurdsson, H., W. Cornell, and S. Carey. 1990. "Influence of Magma Withdrawal on Compositional Gradients During the A.D. 79 Vesuvius Eruption." *Nature* 345: 519–21.

21. Carey, S. N., and H. Sigurdsson. 1986. "The Eruption of Vesuvius in AD 79: II Variations in Column Height and Discharge Rate." *Bull. Geol. Soc. Amer.* 99: 303–14.
22. Sigurdsson et al. 1985. Op. Cit.

Chapter 6

1. With the rise of Christianity the incentive for the study of nature faded. Medieval Christianity led to self-repression and otherworldliness. Man now had a complete theory as to *why* things happened, based on Scripture, and was therefore no longer interested in *how* they happened. As stated in the twelfth century by Peter Lombard of Paris: "Just as man is made for God, in order to serve Him, so the universe is made for man, to serve him. For this reason man was placed in the center of the universe, in order to serve and be served." The Earth had been created ready-made only a few thousand years before and would soon come to an end: "The terror with which men await the end of the world decides me to chronicle the years already passed, that thus one may know exactly how many have elapsed since the Earth began" said St. Gregory the Great (A.D. 540–604). The Christian Church was to react strongly against the blasphemous idea of any knowledge found outside the world of God. Eventually, in the thirteenth century the Church even prohibited the teaching of the Aristotelian cosmology.
2. Boorstin, D. J. 1983. *The Discoverers*. New York: Random House.
3. Turner, A. K. 1993. *The History of Hell*. New York: Harcourt Brace.
4. Kelly, S. 1969. "Theories of the Earth in Renaissance Cosmologies" in C. J. Schneer (ed.), *Toward a History of Geology*. Cambridge: MIT Press, 214–25.
5. Carli, E. 1979. *The Landscape in Art*. New York: William Morrow. Clark, K. 1996. *Landscape into Art*. London: John Murray [see plate 39, Prado Museum, Madrid].
6. Brock, W. H. 1992. *The Norton History of Chemistry*. London: Norton.
7. Gortani, M. 1963. "Italian Pioneers in Geology and Mineralogy." *Cahiers d'Histoire Mondiale* 7: 503–19.
8. Pachter, H. M. 1951. *Magic into Science; The Story of Paracelsus*. New York: Henry Schuman. Paracelsus was in many ways the ultimate medieval alchemist—a legendary "magus" always in search of new knowledge.
9. Rossi, P. 1984. *The Dark Abyss of Time*. Chicago: Univ. of Chicago Press, 338. Read, J. 1961. *Through Alchemy to Chemistry*. London: G. Bell and Sons, 206.

Chapter 7

1. The spectacular voyages of discovery to distant continents and circumnavigations of the Earth contributed greatly to this new mental outlook on the world. This discovery of the Earth extended far beyond the geographical sphere and found its expression, for example, in the loving representation of natural scenery and landscape in art and in the actual study of the Earth. The seeds of the new philosophy were mainly sown in the Italian city-states, and blossomed most of all in Florence. (Goldstein, T. 1980. *Dawn of Modern Science*. Boston: Houghton Mifflin.
2. In the life of Leonardo da Vinci and many other Renaissance men, art and science converged closely and often became indistinguishable. The breakthrough in art into three-dimensional representation of form was spearheaded by the Florentine artist Giotto di Bondone in the fourteenth century and reached its peak in the anatomical and other scientific illustrations of Leonardo da Vinci and the Flemish physician Andreas Vesalius. This new dimension in art had a profound influence on the natural sciences because the graphic representation of objects, minerals, or landforms provided a means of recording observations and transmitting and storing information. Leonardo made good use of the new dimension when he drew his first landscape at the age of twenty-one, a panoramic sketch of the

Arno Valley in 1473 (Goldstein, *Dawn of Modern Science*). Gombrich, E. H. 1953. "Renaissance Artistic Theory and the Development of Landscape Painting." *Gazette des Beaux-Arts* 41: 335–60.

3. Clark, K. 1966. *Landscape into Art*. London: John Murray.
4. The English edition of Agricola's classic work *De Re Metallica* was finally issued in 1912 in the translation of the American president and geologist Herbert Hoover (without doubt the only chief of state who has ever translated one of the great scientific classics). Agricola, G. 1950. *De Re Metallica*, New York: Dover. Sarton, G. 1957. *Six Wings: Men of Science in the Renaissance*. Bloomington: Indiana Univ. Press.
5. La Rocque, A. 1957. *The Admirable Discourses of Bernard Palissy*. Urbana: Univ. of Illinois Press.
6. Most of Kircher's life was spent as a Jesuit priest in Rome, where he studied, lectured, experimented, and wrote on a variety of topics as a professor of the Roman College until his death. Reilly, C. 1955. "Athanasius Kircher S. J., A Contemporary of the 'Sceptical Chymist.' " *J. Chem. Education* 32: 253–58. Reilly, C. 1974. *Athanasius Kircher, S. J.: Master of a Hundred Arts*, vol. I. Rome: Studia Kircheriana.
7. Sapper, K. 1932. "Athanasius Kircher als Geograph" in *Aus der Vergangenheit der Universität Würzburg*. Berlin: Hrsg. Max Buchner, 355–62. Strasser, G. F. 1982. "*Specaculum Vesuvii*: Zu zwei neuentdecten Handschriften von Athanasius Kircher mit seinen Illustrationsvorlagen" in *Theatrum Europaeum, Festschrift für Elida Maria Szarota*. Munich: Wilhelm Fink Verlag, 363–84. Godwin, J. 1979. *Athanasius Kircher: A Renaissance Man and the Quest for Lost Knowledge*. London: Thames and Hudson. The *Mundus Subterraneus* was published in Amsterdam in 1665 and other editions followed in 1668 and 1678. An English version was published in London in 1699 with the title *The Vulcanoes; or Burning and Fire-vomiting Mountains, Famous in the World.*
8. Nicolson, M. H. 1959. *Mountain Gloom and Mountain Glory*. Ithaca, N.Y.: Cornell Univ. Press.
9. Rodolico, F. 1971. "Niels Stensen, Founder of the Geology of Tuscany." *Acta Historica Scientiarum Naturalium et Medicinalium* 23: 237–43.
10. Debus, A. G. 1967. "Fire Analysis and the Elements in the Sixteenth and Seventeenth Centuries." *Annals of Science* 23: 127–47.
11. Winchelsea, Earl. 1669. *A true relation of the late prodigious earthquake and eruption of Mount Aetna, or Monte-Gibello*. T. Newcomb in the Savoy. Rodwell, G. F. 1878. *Etna, a history of the mountain and of its eruptions*. London: C. Kegan Paul & Co. Hyde, W. W. 1916. "The Volcanic History of Etna." *The Geographical Review* 1: 401–18. Chester, D. K., A. M. Duncan, J. E. Guest, and C. R. J. Kilburn. 1985. *Mount Etna*. Stanford: Stanford Univ. Press.
12. Middleton, K. W. E. 1982. "The 1669 Eruption of Mount Etna: Francesco d'Arezzo on the Vitreous Nature of Lava." *Arch. Nat. Hist.* 11: 99–102.
13. The attitude to nature in painting in the Middle Ages was dictated largely by the Christian religion and nature was for the Christian a manifestation of divine power. In this environment, landscape was more the landscape of symbol rather than the depiction of scene and place. Consequently, landscape in the Gothic art of the Middle Ages took the form of backdrop to religious scenes. In terms of geologic expression, however, some interesting landscape depiction can be found in late Gothic scenes devoted to Purgatory and Hell.
14. The artist Domenico Gargiulo (1612–1675) was a native of Naples and was also known by the nickname Micco Spadaro (Mac the Knife), presumably because of his father's profession as a swordsmith. Gargiulo is well known for his realistic landscapes of historic events, dramatically portrayed and successfully conveying effects of atmosphere. Roworth, W. W. 1993. "The Evolution of History Painting: Masaniello's Revolt and Other Disasters in Seventeenth-century Naples." *The Art Bull.* 75: 219–34.
15. Mercalli, G. 1883. *Vulcani e Fenomeni Vulcanici in Italia*. Milan: Arnaldo Forni Editore. Alfano, G. B., and I. Friedlaender. 1929. *Die Geschichte des Vesuv*. Berlin: Dietrich Reimer.

16. In the Constantini Collection, Rome.
17. In the Kunsthistorisches Museum, Vienna.
18. Prior to the 1631 eruption, a volcanic eruption had been depicted only in iconic form, highly schematic, and usually as part of the background stagework of lithographs of the mythological figures of the god Vulcan, or in depictions of Fire as one of the four elements. Thus, in the background of Adriaen Collaert's engraving of fire, several small volcanic cones spew out flames, with a column of smoke rising to the heavens. In this and many other woodcuttings and lithographs, the icon of a volcano reflects the universal view that volcanic eruptions were the consequence of combustion of flammable matter within the earth, such as coal or sulfur. This view persisted towards the end of the eighteenth century, when the studies of Sir William Hamilton and others on Vesuvius showed that volcanic eruption was, in fact, the flow of molten rock from the interior of the earth to the surface.
19. The attribution of this work to D'Angelo may be questioned, however, because in one source he is reported to have died in 1629 or two years before the eruption date. *Civilta del Seicento a Napoli*, vol. 1. Napoli: Electa, 1984.

Chapter 8

1. Whewell, W. 1857. *History of the Inductive Sciences*, vol. III. London: John W. Parker.
2. Carozzi, A. V. 1970. "Robert Hooke, Rudolf Erich Raspe, and the Concept of 'Earthquakes.' " *ISIS* 61: 85–91. Ranalli, G. 1982. "Robert Hooke and the Huttonian theory." *J. Geol.* 90: 319–25.
3. Descartes lived in Holland most of his life and was free of the religious persecution that restricted the practice of science in most Catholic countries: "What other country, where you can enjoy such perfect liberty, where you can sleep with more security, where poisoning, treacheries, calumnies are less known, and where there has survived more of the innocence of our forefathers?"
4. Roger, J. 1962. [Introduction to Buffon] *Les époques de la nature, mémoires du Musée National d'Histoire Naturelle*, Série C. Paris: Sciences de la Terre, 10. Rossi, P. 1984. *The Dark Abyss of Time*. Chicago: Univ. of Chicago Press, 338.
5. Carozzi, A. V. 1968. *De Maillet's Telliamed*. Urbana: Univ. of Illinois Press.
6. Rossi, *The Dark Abyss of Time*, 338.

Chapter 9

1. La Rocque, A. 1957. *The Admirable Discourses of Bernard Palissy*. Urbana: Univ. of Illinois Press.
2. Challinor, J. 1953. "The Early Progress in British Geology." *Annals of Science* 9: 124–53.
3. Some of the cliffs contain truly enormous columns, such as those at Fairhead, up to 100 m in height and more than 1 m in diameter. One of the earliest notes of columnar basalt in Britain was by Edward Lhwyd, Keeper of the Museum at Oxford (1695), who described the rocks of Glyder Mountain, North Wales, as "adorn'd with equidistant Pillars, and these again slightly cross'd at certain joynts." Lhwyd also visited the Giant's Causeway in 1699. Challinor, J. 1953. "The Early Progress in British Geology." *Annals of Science* 9: 124–53. De la Beche, H. T. 1837. *Researches in Theoretical Geology*. New York. Porter, R. 1977. *The Making of Geology*. Cambridge: Cambridge University Press. Tomkeieff, S. I. 1940. "The Basalt Lavas of the Giant's Causeway District of Northern Ireland." *Bull. Volcan.* 6: 89–144.
4. Anglesea, M., and J. Preston. 1980. "'A Philosophical Landscape' Susanna Drury and the Giant's Causeway." *Art Hist.* 3: 252–73.
5. Gillispie, C. C. 1959. *Genesis and Geology*. New York: Harper & Row.
6. Geikie, A. 1897. *Ancient Volcanoes of Great Britain*, 2 vols. London.
7. Smith, C. S. 1969. "Porcelain and Plutonism" in C. J. Schneer (ed.), *Toward a History of Geology*. Cambridge: MIT Press, 317–38.

8. Richard Kirwan was president of the Royal Irish Academy from 1799 to 1819 and a well-known scientific figure, recognized mineralogist, and His Majesty's Inspector of Mines in Ireland.
9. Sweet, J. M., and C. D. Waterston. 1967. "Robert Jameson's Approach to the Wernerian Theory of the Earth." *Annals of Science* 23: 81–95.
10. Darwin, F. 1887. *Life and Letters of Charles Darwin*, vol. 1. London.
11. Baumgartel, H. 1969. "Alexander von Humboldt: Remarks on the Meaning of Hypothesis in his Geological Researches" in Schneer, *History of Geology*, 19–35.
12. Adams, F. D. 1938. *The Birth and Development of the Geological Sciences*. New York: Dover.
13. Goethe, J. W. 1985. *Italian Journey*. Harmondsworth, England: Penguin.
14. Dolomieu was an officer in the French army and traveled widely in volcanic regions in Sicily and central Italy. He became professor at the Paris School of Mines, and participated in the French Expedition of Napoleon to Egypt.
15. Greene, M. T. 1982. *Geology in the Nineteenth Century*. Ithaca: Cornell Univ. Press, 324.
16. Nieuwenkamp, W. 1970. "Leopold von Buch 1774–1853." 552–57. *Dictionary of Scientific Biography* II. Nieuwenkamp, W. 1975. "Trends in Nineteenth Century Petrology." *Janus* 62: 235–69.

Chapter 10

1. Porter, R. 1977. *The Making of Geology*. Cambridge: Cambridge Univ. Press.
2. Sleep, M. C. W. 1969. "The Geological Work of Sir William Hamilton." *Proceedings of the Geologists' Association* 80: 353–63.
3. Childe-Pemberton, W. S. 1924. *The Earl Bishop*. London: Hurst and Blackett.
4. Debus, A. G. 1969. "Edward Jorden and the Fermentation of the Metals" in Schneer, *History of Geology*, 100–21.
5. Faul, H., and C. Faul. 1983. *It Began with a Stone*. New York: Wiley.
6. Beer, G. de. 1962. "The Volcanoes of Auvergne." *Annals of Science* 18: 49–61.
7. Scrope, G. P. 1825. *Considerations on volcanos, the probable causes of their phenomena, the laws which determine their march, the disposition of their products, and their connexion with the present state and past history of the globe; leading to the establishment of a new theory of the Earth*. London: W. Phillips.
8. Fenton, C. L., and M. A. Fenton. 1952. *Giants of Geology*. New York: Doubleday.
9. Geikie, A. 1905. *The Founders of Geology*. New York: Dover, 486.
   McKie, D. 1952. *Antoine Lavoisier*. New York: Henry Schuman.
10. Briggs, J. M. 1989. "Antoine de Genssane, Mining, Vapors and Mineralogy in Eighteenth Century France." *EOS*: 1511–24.
11. Spallanzani, L. Abbé. 1792–97. *Travels in the Two Sicilies and Some Parts of the Apennines*, 4 vols. London.
12. Burke, J. 1978. *Connections*. Boston: Little Brown.

Chapter 11

1. Carswell, J. 1950. *The Prospector, Being the Life and Times of Rudolf Erich Raspe (1737–1794)*. London: Cresset Press.
2. Haarmann, E. 1942. "Ein Münchhausen als Geologe: Rudolf Erich Raspe, 1736–1794." *Geologische Rundschau*, 33: 104–20.
3. Hallo, R. 1934. *Rudolf Erich Raspe, Ein Wegbereiter von Deutscher Art und Kunst*. Stuttgart: Verlag W. Kohlhammer.
4. Carozzi, A. V. 1969. "Rudolf Erich Raspe and the Basalt Controversy." *Studies in Romanticism* 8: 235–50.
5. Cerswell, *The Prospector*.

6. Gerstner, P. A. 1968. "James Hutton's Theory of the Earth and his Theory of Matter." *Isis* 59: 26–31. Gerstner, P. A. 1971. "The Reaction to James Hutton's Use of Heat as a Geological Agent." *Brit. J. Hist. Sci.* 5: 353–62. Donovan, A. 1978. "James Hutton, Joseph Black and the Chemical Theory of Heat." *Ambix* 25: 176–90.
7. Gillispie, C. C. 1959. *Genesis and Geology*. New York: Harper and Row.
8. Gerstner 1971. Op. Cit.
9. Eyles, V. A. 1963. "The Evolution of a Chemist." *Annals of Science* 19: 153–182.
10. Eyles, V. A. 1961. "Sir James Hall, Bt. (1761–1832)." *Endeavour*: 210–16. Eyles, "The Evolution of a Chemist." Smith, C. S. 1969. "Porcelain and Plutonism" in Schneer, *History of Geology*, 317–38. Yoder, H. S. 1980. "Experimental Mineralogy: Achievements and Prospects." *Bull. Mineralogie* 103: 5–26. Hall's hypothesis was apparently prompted by the accidental spill of a large volume of silicate melt at Leith glassworks near Edinburgh. The spilled glass had cooled extremely slowly and produced a granular and crystalline solid resembling rock. A similar glass-factory accident was reported in 1809, at a meeting of the Literary and Philosophical Society of Newcastle: "Mr. Cookson produced a piece of glass which had been allowed to cool very slowly and so exhibited the opacity and other properties of a crystallized stony body. In illustration of this phenomenon, Dr. Townson requested that an extract might be read from Sir James Hall's paper on Whin and Lava. This phenomenon has since been exhibited on rather too large a scale by the falling in of the dome of the furnace of Mr Cookson's bottle house in the Close. Some tons of the melted glass had in consequence of the accident an opportunity of cooling gradually and exhibited all the varieties between perfect glass and an opake crystallized substance" (Porter, R. 1977. *The Making of Geology*. Cambridge: Cambridge University Press).
11. One of the reasons for Hall's belief that carbonate minerals would fuse under pressure was his observation of "nodules of calcareous spar inclosed in whinstone" (Hall, J. 1805. "Experiments on Whinstone and Lava." *Trans. Roy. Soc. Edinburgh* 5: 43–75; Hall, J. 1812. "Account of a Series of Experiments, Shewing the Effects of Compression in Modifying the Action of Heat." *Trans. Roy. Soc. Edinburgh* 6: 71–187). Rounded calcite nodules or amygdules are quite common in basalt lavas and minor intrusions and Hall interpreted them as having formerly been liquid during the eruption or intrusion of the magma: "the whin and the spar had been liquid together; the two fluids keeping separate like oil and water." We now know, of course, that such calcite amygdules in lavas are secondary in origin—that is, they are the infillings of gas cavities in the lava long after its solidification from magma.
12. When all his accomplishments are considered, Hall must be regarded as the true founder of experimental petrology. It is therefore surprising that Eugster (Eugster, H. P. 1971. "The Beginnings of Experimental Petrology." *Science* 173: 481–89) has claimed that the work of Van't Hoff in 1896 to 1908 on the phase relations of marine evaporites marks the beginnings of experimental petrology. As the above examples of the work of James Hall clearly demonstrate, the beginnings of experimental petrology precedes Van't Hoff's work by at least one century.
13. Yoder, H. S. 1980. "Experimental Mineralogy: Achievements and Prospects." *Bull. Mineralogie* 103: 5–26.
14. Smith, C. S. 1969. "Porcelain and Plutonism" in Schneer, *History of Geology*.

Chapter 12

1. Davidson, C. 1927. *The Founders of Seismology*. Cambridge: Cambridge Univ. Press.
2. Michell, J. 1760. "The Nature and Origin of Earthquakes." *Phil. Trans. Roy. Soc. London* LI:566–74.
3. Reinhardt, O., and D. R. Oldroyd. 1983. "Kant's Theory of Earthquakes and Volcanic Action." *Annals of Science* 40: 247–72.
4. Lawrence, P. 1978. "Charles Lyell versus the Theory of Central Heat: A Reappraisal of Lyell's Place

in the History of Geology." *Journal of the History of Biology* 11: 101–28.
5. Reinhardt, O., and D. R. Oldroyd. 1984. "By Analogy with the Heavens: Kant's Theory of the Earth." *Annals of Science* 41: 203–21.
6. Siegfried, R., and R. H. Dott, Jr. 1976. "Humphry Davy as Geologist, 1805–29." *British Journal of the History of Science* 9: 219–27.
7. Davy, H. 1839–40. *The Collected Works of Sir Humphry Davy*, 9 vols. J. Davy (ed.). London: Smith, Elder.
8. Paris, J. A. 1830. *The Life of Sir Humphry Davy*. London.
9. Siegfried and Dott, "Humphry Davy."
10. Davy, H. 1840. *Collected Works*, vol. VI, ch. XX, 344–58. Treneer, A. 1963. *The Mercurial Chemist*. London: Methuen.
11. Anon. 1835. "Ampère's Theory of the Formation of the Globe, and of the Phenomena of Volcanoes." *The Edinburgh Philosophical Journal* 18: 339–47.
12. Bischof, G. 1839. "On the Natural History of Volcanos and Earthquakes." *The Edinburgh New Philosophical Journal* 26: 25–81; 347–86.
13. Gay-Lussac, L. J. 1823. "Reflexions sur les Volcans." *Annales de Chimie et de Physique* 415–29.
14. Bischof, "Natural History of Volcanoes."
15. Mallet, R. 1873. "Volcanic Energy: An Attempt to Develop Its True Origin and Cosmical Relations." *Phil. Trans. Roy. Soc. London* 163: 147–227.
16. Bischof, "Natural History of Volcanos." Bischof, G. 1841. *Leonhard und Bonn's Neues Jahrbuch für Mineralogie*, 565–66. Bischof, G. 1841. *Physical, Chemical and Geological Researches on the Internal Heat of the Globe*. London: Longmans. Daubney, C. 1839. "Reply to Professor Bischof's Objections to the Chemical Theory of Volcanos." *The Edinburgh New Philosophical Journal* 26: 291–99.
17. Lyell, C. 1835. *Principles of Geology*. London: James Murray.
18. Smith, C. S. 1989. "William Hopkins and the Shaping of Dynamical Geology: 1830–1860." *Brit. J. Hist. Sci.* 22: 27–52.
19. Lawrence, "Charles Lyell Versus the Theory of Central Heat."
20. Lyell. *Principles of Geology*.
21. Whewell, W. 1837. *History of the Inductive Sciences, from the Earliest to the Present Times*. London.
22. Anon. 1875. "The Internal Heat of the Earth." *Nature* 12: 545–46.
23. Judd, J. 1881. *Volcanoes: What They Are and What They Teach*. New York: D. Appleton and Co.
24. Thomson, W. 1889. *Popular Lectures and Addresses*, 3 vols. London: Macmillan.
25. Bonney, T. G. 1899. *Volcanoes, Their Structure and Significance*. New York: G. P. Putnum's Sons.
26. Brock, W. H. 1979. "Chemical Geology or Geological Chemistry?" in L. J. Jordanova and R. S. Porter (eds.). *Images of the Earth. Brit. Soc. Hist. Sci. Monograph* 1: 147–70.
27. Day, A. L. 1925. "Some Causes of Volcanic Activity." *J. Franklin Inst.* 200: 161–82.
28. Jeffreys, H. 1952. *The Earth*. Cambridge: Cambridge Univ. Press.

Chapter 13

1. Brush, S. G. 1979. "Nineteenth-century Debates About the Inside of the Earth: Solid, Liquid or Gas?" *Annals of Science* 36: 225–54.
2. Brush, S. G. 1976. *The Kind of Motion We Call Heat*. North-Holland: Amsterdam.
3. Spallanzani, L. Abbé. 1792–97. *Travels in the Two Sicilies and Some Parts of the Apennines*, 4 vols. London.
4. Scrope, G. P. 1825. *Considerations on volcanos, the probable causes of their phenomena, the laws which determine their march, the disposition of their products, and their connexion with the present state and past history of the globe; leading to the establishment of a new theory of the Earth*. London: W. Phillips.

5. Kircher, A. 1664. *Mundus Subterraneus*, 2 vols. Amsterdam. Figuier, L. 1870. *Earth and Sea*. Translated by W. H. D. Adams. London: Nelson and Sons.
6. Arago, F. 1839. "Historical Eloge of Joseph Fourier." *The Edinburgh New Philosophical Journal*, 26: 1–25; 217–44.
7. Fourier, J. B. J. 1822. *Théorie analytique de la chaleur*. Paris.
8. Brush, *The Kind of Motion We Call Heat*.
9. Fourier, J. B. J. 1836. "General Remarks on the Temperature of the Terrestrial Globe and the Planetary Species." *Amer. J. Science* 32: 1–20.
10. Cooper, T. 1829. "Analysis of an 'Essai sur la temperature de l'interieur de la terre, par M. Cordier.'" *Amer. J. Science* 15: 109–30.
11. Poisson, S. D. 1835. *Théorie mathématique de la chaleur*. Paris: Bachelier. Poisson, S. D. 1837. "Mémorie sur les températures de la partie solide du globe." *Compte Rendu des Séances de l'Academie des Sciences* 4: 137–66.
12. Franklin, B. 1793. "Conjectures Concerning the Formation of the Earth, etc. in a Letter to the Abbé Soulavie." *Trans. Amer. Phil. Soc.* 3: 1–5.
13. Sigurdsson, H. 1982. "Volcanic Pollution and Climate: The 1783 Laki Eruption." *EOS* 63: 601–02.
14. Eaton, A. 1830. *Geological Textbook*. Albany.
15. Darwin, C. 1840. "On the Connexion of Certain Volcanic Phenomena in South America; and on the Formation of Mountain Chains and Volcanos, as the Effect of the Same Power by Which Continents Are Elevated." *Transactions of the Geological Society of London* 5: 601–31.
16. Herschel, J. 1837. *Proc. Geol. Soc. London* 2: 596–601. [extract from a letter]
17. Mallet, R. 1858. "Fourth Report upon the Facts and Theory of Earthquake Phenomena." London: *Report of the British Association*, 1–136. Mallet, R. and J. W. Mallet. 1858. *The Earthquake Catalogue of the British Association*. London: Taylor and Francis. Mallet, R. 1873. "Volcanic Energy: An Attempt to Develop its True Origin and Cosmical Relations." *Phil. Trans. Roy. Soc. London* 163: 147–227.
18. Dana, J. D. 1847. "Geological Results of the Earth's Contraction in Consequence of Cooling." *Amer. J. Sci.* 3: 176–88. Dana, J. D. 1847. "A General Review of the Geological Effects of the Earth's Cooling from a State of Igneous Fusion." *Amer. J. Science* 4: 88–92. Belli, G. 1850. "Pensiri sulla consistenza ecc. della crosta solida terrestre." *Journals of the Institute of Lombardy*, vol. II.
19. Dana, J. D. 1873. "On Some Results of the Earth's Contraction from Cooling, Including a Discussion of the Origin of Mountains, and the Nature of the Earth's Interior." *Amer. J. Science* 5: 423–43. Dana, J. D. 1873. "On Some Results of the Earth's Contraction from Cooling, Part II, The Condition of the Earth's Interior." *Amer. J. Science* 6: 6–14; 104–15. Dana, J. D. 1873. "On Some Results of the Earth's Contraction from Cooling, Including a Discussion of the Origin of Mountains." *Phil. Mag.* 46: 41–219. Hopkins, W. 1839. "On the Phenomena of Precession and Nutation, Assuming Fluidity of the Interior of the Earth." *Phil. Trans. Roy. Soc. London*: 381–423. Hopkins, W. 1840. "On Precession and Nutation, Assuming the Interior of the Earth to be Fluid and Heterogeneous." *Phil. Trans. Roy. Soc. London*: 193–208. Hopkins, W. 1842. "Researches in Physical Geology—Third Series. On the Thickness and Constitution of the Earth's Crust." *Phil. Trans. Roy. Soc. London* 132: 43–45.
20. Perhaps the earliest use of the magnetic needle to aid navigation was in China in about A.D. 1000. In Europe, however, the magnetic needle was not known until two centuries later and was documented by the English monk Alexander Neckam (1157–1217). Its popularity grew quickly and by the late thirteenth century the magnetic needle was in general use by navigators throughout the Mediterranean.
21. Sarton, G. 1957. *Six Wings: Men of Science in the Renaissance*. Bloomington: Indiana Univ. Press.
22. Newton's estimate of the Earth's density was in fact highly accurate. In 1775 Nevil Maskelyne, the Astronomer Royal of Britain, had inferred that the density was about 4.5 times that of water on the

basis of pendulum measurements, but in 1798 the English scientist Henry Cavendish, using a much more accurate torsion balance built by Cambridge geologist John Michell, measured the gravitational attraction between two bodies of known masses (the gravitational constant) and used this to "weigh the Earth." The average density of the Earth, he calculated, was 5488 times greater than that of water, a value remarkably close to the present-day accepted value of 5.517 g/cm.^{3} This result strongly implied the presence of a very dense, metallic material within the Earth, even heavier than basalt, the densest known rock at the surface, which has a density of about 2.85 g/cm.^{3} The early explanations of high density for the Earth relied entirely on the effect of composition, rather than pressure, on the density of rocks at great depth.

23. Kushner, D. S. 1990. "The Emergence of Geophysics in Nineteenth Century Britain." Ph.D. thesis, Princeton University.
24. Hopkins, W. 1839. "On the Phenomena of Precession and Nutation, Assuming Fluidity of the Interior of the Earth." *Phil. Trans. Roy. Soc. London*: 381–423. Hopkins, W. 1840. "On Precession and Nutation, Assuming the Interior of the Earth to be Fluid and Heterogeneous." *Phil. Trans. Roy. Soc. London* 193–208. Hopkins, "Researches in Physical Geology."
25. Smith, C. S., and M. N. Wise. 1989. *Energy and Empire, A Biographical Study of Lord Kelvin*. Cambridge: Cambridge Univ. Press.
26. Fisher, O. 1872. "On the Rigidity of the Earth and the Liquidity of Lavas." *Nature* 6: 241.
27. Fisher, O. 1881. *Physics of the Earth's Crust*. London: Macmillan.
28. Kelvin, W. T. 1863. "On the Secular Cooling of the Earth." *Phil. Mag.* 25: 1–14.
29. Delaunay, M. 1868. "On the Hypothesis of the Internal Fluidity of the Terrestrial Globe." *Geol. Mag.* 5: 507–11.
30. Forbes, D. 1870. "On Volcanos." *Geol. Mag.* 7: 314–28.
31. Pratt, J. H. 1871. "The Solid Crust of the Earth Cannot be Thin." *Phil. Mag.* 42: 280–90.
32. The full title of Scrope's work is *Volcanos, the character of their phenomena, their share in the structure and composition of the surface of the globe, and their relation to its internal forces. With a descriptive catalogue of all known volcanos and volcanic formations*. Published in 1862 (second ed., 1872).
33. Burchfield, J. D. 1975. *Lord Kelvin and The Age of the Earth*. Chicago: University of Chicago Press.
34. Helmholz, H. 1856. "On the Interaction of Natural Forces." *Phil. Mag.* 11: 489–518.
35. Smith, C. S. 1989. William Hopkins and the Shaping of Dynamical Geology: 1830–1860." *Brit. J. Hist. Sci.* 22: 27–52.
36. Brush, *The Kind of Motion We Call Heat*.
37. Barus, C. 1893. "High-temperature Work in Igneous Fusion and Ebullition Chiefly in Relation to Pressure." *U.S. Geol. Surv. Bull.* 103: 57. Barus, C. 1893. "The Fusion Constants of Igneous Rocks. Part 3, the Thermal Capacity of Igneous Rock." *Phil. Mag.* 35: 296–307. Barus, C. 1893. "The Contraction of Molten Igneous Rock on Passing from Liquid to Solid." *U.S. Geol. Surv. Bull.* 103: 25–55. Barus, C. 1893. "Mr. McGee and the Washington Symposium." *Science* 22: 22–23. Barus, C. 1893. "Criticism of Mr. Fisher's Remarks on Rock Fusion." *Amer. J. Science* 146: 140–41. Barus, C. 1893. "Isothermals and Isometrics Relative to Viscosity." *Amer. J. Science* 45: 87–96. King, C. 1893. "The Age of the Earth." *Amer. J. Sci.* 45, 265: 1–20.
38. Richter, F. M. 1986. "Kelvin and the age of the Earth." *J. Geology* 94: 395–401.
39. Harrison, T. M. 1987. "Comment on 'Kelvin and the age of the Earth.' " J. *Geology* 95: 725–27.
40. Kushner, "Geophysics."
41. Darwin, G. H. 1878. "On the Influence of Geological Changes on the Earth's Axis of Rotation." *Phil. Trans. Roy. Soc. London*, 167: 271–312.
42. Darwin, G. H. 1882. "On the Stresses Caused in the Interior of the Earth by the Weight of Continents and Mountains." *Phil. Trans. Roy. Soc. London* 173: 187–230.

Chapter 14

1. Brown, S. C. 1979. *Benjamin Thompson, Count Rumford.* Cambridge: MIT Press.
2. Playfair, J. 1802. *Illustrations of the Huttonian Theory of the Earth.* Edinburgh: William Creech.
3. Eyles, V. A. 1961. "Sir James Hall, Bt. (1761–1832)." *Endeavour*: 210–16. Donovan, A. 1978. "James Hutton, Joseph Black and the Chemical Theory of Heat." *Ambix* 25: 176–90.
4. Opie, R. 1929. "A Neglected English Economist: George Poulett Scrope." *The Quarterly Journal of Economics* 44: 101–37. Sturges, P. 1984. *A Bibliography of George Poulett Scrope, Geologist, Economist and Local Historian.* Boston: Baker Library, Harvard Business School.
5. Scrope, G. P. 1825. *Considerations on volcanos, the probable causes of their phenomena, the laws which determine their march, the disposition of their products, and their connexion with the present state and past history of the globe; leading to the establishment of a new theory of the Earth.* London: W. Phillips.
6. Sturges, *Bibliography of Scrope.*
7. Scrope, G. P. 1875. "Notes on the Volcanic Eruptions in Iceland." *Geol. Mag.* 12: 289–91.
8. Scrope, *Considerations.*
9. Scrope, G. P. 1830. "Principles of geology...by Charles Lyell" (review of the 1st ed.) *The Quarterly Review* 43: 411–69. Scrope, G. P. 1835. "Principles of geology...by Charles Lyell" (review of the 3rd ed.) *The Quarterly Review* 53: 406–48.
10. Scrope, G. P. 1862; 1872. *Volcanos, the character of their phenomena, their share in the structure and composition of the surface of the globe, and their relation to its internal forces. With a descriptive catalogue of all known volcanos and volcanic formations.* London: Longman, Green.
11. Scrope, G. P. 1868. "Some Observations on the Supposed Internal Fluidity of the Earth." *Geol. Mag.* 5: 537–39.
12. Hopkins, W. 1839. "On the Phenomena of Precession and Nutation, Assuming Fluidity of the Interior of the Earth." *Phil. Trans. Roy. Soc. London*: 381–423.
13. Ibid.
14. Poisson, S. D. 1835. *Théorie mathématique de la chaleur.* Paris: Bachelier.
    Poisson, S. D. 1837. "Mémorie sur les températures de la partie solide du globe." *Compte Rendu des Séances de l'Academie des Sciences* 4: 137–66.
15. Kushner, D. S. 1990. "The Emergence of Geophysics in Nineteenth Century Britain." Ph.D. thesis, Princeton University.
16. Pole, W. (ed.) 1877. *The Life of Sir William Fairbairn.* London: Longmans.
17. Thomson, J. 1849. "Theoretical Considerations on the Effect of Pressure in Lowering the Freezing Point of Water." *Trans. Roy. Soc. Edinburgh* 16: 575–80. Smith, C. S. and M. N. Wise. 1989. *Energy and Empire, a Biographical Study of Lord Kelvin.* Cambridge: Cambridge Univ. Press.
18. Thomson, W. 1851. "The Effect of Pressure in Lowering the Freezing-Point of Water." *Proc. Roy. Soc. Edinburgh* 2: 267–71.
19. Bischof, G. 1841. *Leonhard und Bonn's Neues Jahrbuch für Mineralogie*: 565–66. Bischof, G. 1841. *Physical, Chemical and Geological Researches on the Internal Heat of the Globe.* London: Longmans. Bischof, G. 1843. *Neues Jahrbuch für Mineralogie*: 1–54.
20. Bunsen, R. 1850. "Uber den Einfluss der Drucks auf die Chemische Natur der Plutonischen Gesteine." *Annalen der Physik und Chemie* 81: 562–67.
21. von Waltershausen, S. 1853. *Uber die Vulkanischen Gesteine in Sicilien und Island und ihre Submarine Umbildung.* Göttingen.
22. Pole, *Sir William Fairbairn.*
23. Hopkins, W. 1854. "Presidential Address." *British Assoc. Adv. Science Report* 23:li–lii. Hopkins, W. 1855. "An Account of Some Experiments on the Effect of Pressure on the Temperature of Fusion of

Different Substances." *British Assoc. Adv. Science* (24th meeting) 2: 57–58.

24. Mallet, R. 1873. "Volcanic Energy: An Attempt to Develop Its True Origin and Cosmical Relations." *Phil. Trans. Roy. Soc. London* 163: 147–227. Nies, F. 1888. *Ueber das Verhalten der Silicate beim Uebergange aus dem gluthflussigen in den festen Aggregatzustand. Programm zur 70. Jahresfeier der K. Wurttemb. Landw. Academie Hohenheim.* Stuttgart: E. Koch.
25. Judd, J. 1881. *Volcanoes: What They Are and What They Teach.* New York: D. Appleton and Co.
26. King, C. 1878. *U.S. Geological Exploration of the Fortieth Parallel*, vol. I, *Systematic Geology*. Washington.
27. Lobley, L. J. 1889. *Mount Vesuvius.* London: Roper and Drowley.
28. Geikie, A. 1897. *Ancient Volcanoes of Great Britain*, 2 vols. London.
29. Darwin, C. 1840. "On the Connexion of Certain Volcanic Phenomena in South America; and on the Formation of Mountain Chains and Volcanos, as the Effect of the Same Power by Which Continents Are Elevated." *Trans. Geol. Soc. London* 5: 601–31. Dana, J. D. 1847. "Geological Results of the Earth's Contraction in Consequence of Cooling." *Amer. J. Science* 3: 176–88. Dana, J. D. 1847. "A General Review of the Geological Effects of the Earth's Cooling from a State of Igneous Fusion." *Amer. J. Science* 4: 88–92.
30. Roberts-Austen, W. C., and A. W. Rücker. 1891. "On the Specific Heat of Basalt." *Phil. Mag.* 32: 353–55.
31. Becker, G. F. 1905. "Introduction" in *The Isomorphism and Thermal Properties of the Feldspars*. Washington, D.C.: Carnegie Institution, 5–12.
32. Barus, C. 1891. "The Contraction of Molten Rock." *Amer. J. Science* 142: 498–99. Barus, C. 1892. "The Relation of Melting Point to Pressure in Case of Igneous Rock Fusion." *Amer. J. Science* 43: 56–57.
33. Barus, C. 1893. "High-temperature Work in Igneous Fusion and Ebullition Chiefly in Relation to Pressure." *U.S. Geol. Surv. Bull.* 103. Barus, C. 1893. "The Fusion Constants of Igneous Rocks. Part 3, the Thermal Capacity of Igneous Rock." *Phil. Mag.* 35: 296–307. Barus, C. 1893. "The Contraction of Molten Igneous Rock on Passing from Liquid to Solid." *U.S. Geol. Surv. Bull.* 103: 25–55. King, C. 1893. "The Age of the Earth." *Amer. J. Science* 45, 265: 1–20.
34. Arrhenius, S. 1900. "Zur Physik des Vulkanismus." *Geol. Foren. Forhandl.* 22: 395–419.
35. Harker, A. 1909. *The Natural History of Igneous Rocks.* New York: Macmillan.
36. Bridgman, P. W. 1914. "Change of Phase Under Pressure." *The Physical Review* 3: 153–203.
37. Daly, R. A. 1933. *Igneous Rocks and the Depths of the Earth.* New York: McGraw-Hill.
38. Shand, S. J. 1949. *Eruptive Rocks.* New York: Wiley.

Chapter 15

1. Liebenow, C. H. 1904. "Notiz uber die Radiummenge der Erde." *Physikalische Zeitschrift* 5: 625–26.
2. Holmes, A. 1915. "Radioactivity and the Earth's Thermal History. Part II. Radio-activity and the Earth as a Cooling Body." *Geol. Mag.*: 102–12.
3. Verhoogen, J. 1980. *Energetics of the Earth.* Washington, D.C.: National Academy of Sciences, 139.
4. Stacey, F. D. 1977. *Physics of the Earth.* New York: Wiley.
5. Hopkins, W. 1839. "On the Phenomena of Precession and Nutation, Assuming Fluidity of the Interior of the Earth." *Phil. Trans. Roy. Soc. London*: 381–423. Fisher, O. 1881. *Physics of the Earth's Crust.* London: Macmillan.
6. Crook, T. 1933. *History of the Theory of Ore Deposits.* London: T. Murby.
7. Loewinson-Lessing, F. Y. 1954. *A Historical Survey of Petrology.* London: Oliver and Boyd.
8. Spallanzani, L. Abbé. 1792–97. *Travels in the Two Sicilies and Some Parts of the Apennines*, 4 vols. London.

9. Kennedy, R. 1805. "A Chemical Analysis of Three Species of Whinstone and Two of Lava." *Trans. Roy. Soc. Edinburgh* 5: 76–98.
10. Darwin, C. 1844. *Geological Observations on the Volcanic Islands*. London. Appleton.
11. Bunsen, R. 1851. "Ueber die Processe der vulkanischen Gesteinsbildungen Islands." *Annalen der Physik und Chemie* 83: 197–272.
12. von Waltershausen, S. 1853. *Uber die vulkanischen Gesteine in Sicilien und Island und ihre submarine Umbildung*. Göttingen.
13. Bunsen, R. 1861. "Uber die Bildung der Granites." *Zeit. Deutsch. Geol. Gesellsch.* 13: 61–63.
14. Forbes, D. 1871. "On the Nature of the Earth's Interior." *Geol. Mag.* 8: 162–73.
15. Bonney, T. G. 1899. *Volcanoes, Their Structure and Significance*. New York: G. P. Putnam's Sons.
16. Clarke, F. W. 1889. "The Relative Abundance of the Chemical Elements." *Bull. Phil. Soc. Wash.* 11: 131–42.
17. Yoder, H. S. 1980. "Experimental Mineralogy: Achievements and Prospects." *Bull. Mineralogie* 103: 5–26. Bowen, N. L. 1913. "The Melting Phenomena of the Plagioclase Feldspars." *Amer. J. Science* 35: 577–99. Bowen, N. L. 1913. "The Later Stages of the Evolution of the Igneous Rocks." *J. Geol.* 23: 1–91.

Chapter 16

1. Darwin, C. 1844. *Geological Observations on the Volcanic Islands*. London: Appleton.
2. Renard, A. 1882. "On the Petrology of St. Paul's Rocks." Appendix B, Narrative of the Challenger Report, vol. 2.
3. Melson, W. G., S. R. Hart, and G. Thompson. 1971. "St. Paul's Rocks, Equatorial Atlantic." Woods Hole Oceanographic Institution Technical Report 71–20.
4. One of the first to propose a peridote source was A. Daubrée in 1879; a similar view was also taken by H. Rosenbusch in 1889 ("Uber die chemischen Beziehungen der Eruptivgesteine." *Min. u. petr. Mitth.*, Vienna, vol. 11).
5. F. D. Adams and E. G. Coker were the first to measure the critical physical properties of a variety of rocks under high pressure and temperature in 1906 ("An Investigation into the Elastic Constants of Rocks, More Especially with Reference to Cubic Compressibility." *Amer. J. Science* 22: 95–123). Later Leason H. Adams pointed out the similarity between the seismologic evidence and the experiments in 1924 ("Temperatures at Moderate Depths Within the Earth." *J. Wash. Acad. Sci.* 14: 459–72).
6. Bowen, N. L. 1928. *The Evolution of the Igneous Rocks*. Princeton, N.J.: Princeton Univ. Press.
7. Rittmann, A. 1960. *Vulkane und ihre Tütigkeit*. Stuttgart: Ferdinand Enke Verlag.
8. Barth, T. F. W. 1962. *Theoretical Petrology*. New York: Wiley.
9. Turner, F. J., and J. Verhoogen. 1960. *Igneous and Metamorphic Petrology*. New York: McGraw-Hill, 694.
10. MacCurdy, E. (ed.). 1939. *The Notebooks of Leonardo da Vinci*. New York: Braziller.
11. Hooke, R. 1665. *Micrographica*.
12. Reinhardt, O., and D. R. Oldroyd. 1984. "By Analogy with the Heavens: Kant's Theory of the Earth." *Annals of Science* 41: 203–21.
13. French, B. 1977. *The Moon Book*. Harmondsworth, England: Penguin.
14. Spallanzani, L. Abbé. 1792–97. *Travels in the Two Sicilies and Some Parts of the Apennines*, 4 vols. London.
15. Breislak, S. 1811. *Introduzione alla Geologia*.
16. Scrope, G. P. 1825. *Considerations on volcanos, the probable causes of their phenomena, the laws which determine their march, the disposition of their products, and their connexion with the present state and past*

history of the globe; leading to the establishment of a new theory of the Earth. London: W. Phillips.

17. Scrope, G. P. 1826; 1878. *The Geology and Extinct Volcanos of Central France*. London: J. Murray. (Reprint; New York: Arno Press, 1978. Scrope, G. P. 1862; 1872. *Volcanos, the character of their phenomena, their share in the structure and composition of the surface of the globe, and their relation to its internal forces. With a descriptive catalogue of all known volcanos and volcanic formations*. London: Longmans Green.
18. Fisher, O. 1881. *Physics of the Earth's Crust*. London: Macmillan.
19. Judd, J. 1881. *Volcanoes: What They Are and What They Teach*. New York: D. Appleton and Co.
20. Bonney, T. G. 1899. *Volcanoes, Their Structure and Significance*. New York: G. P. Putnam's Sons.
21. von Waltershausen, S. 1853. *Uber die Vulkanischen Gesteine in Sicilien und Island und ihre Submarine Umbildung*. Göttingen.
22. Bunsen, R. 1851. "Ueber die Processe der Vulkanischen Gesteinsbildungen Islands." *Annal. Phys. Chemie* 83: 197–272.
23. Johnson-Lavis, H. J. 1891. *The South Italian Volcanoes*. Naples, 43.
24. Tyrrell, G. W. 1931. *Volcanoes*. New York: Holt.
25. Fenner, C. N. 1908. "Features Indicative of Physiographic Conditions Prevailing at the Time of the Trap Extrusions in New Jersey." *J. Geol.* 16: 299–327.
26. Harker, A. 1909. *The Natural History of Igneous Rocks*. New York: Macmillan.
27. Mallet, R. 1858. "Fourth Report upon the Facts and Theory of Earthquake Phenomena." London: Report of the British Association, 1–136. Dean, D. R. 1991. "Robert Mallet and the Founding of Seismology." *Annals of Science* 48: 39–67.
28. Hallam, A. 1983. *Great Geological Controversies*. Oxford: Oxford University Press.
29. The amazing fit of the continents on either side of the Atlantic had, however, been pointed out much earlier—first by Sir Francis Bacon in 1620 and later by Antonio Snider Pelligrini who drafted a map, published in 1858, showing the reconstruction of the landmasses before continental drift.
30. Holmes, A. 1929. "Radioactivity and Earth Movements." *Trans. Geol. Soc. Glasgow* 18: 559–605.
31. Holmes, A. 1915. "Radioactivity and the Earth's Thermal History. Part II. Radio-activity and the Earth as a Cooling Body." *Geol. Mag.* 102–12.
32. Menard, H. W. 1986. *The Ocean of Truth*. Princeton, N.J.: Princeton University Press.
33. A. S. McEwan, et al. 1998. "High-Temperature Silicate Volcanism on Jupiter's Moon Io." *Science*, 281, 87–90.
34. S. J. Peale, P. Cassen, and R. T. Reynolds. 1979. "Melting of Io by tidal dissipation." *Science*, 203, 892–894.

Glossary of Volcanic Terms

Between some of the rocks with different names, there is about as much difference as between Tweedledee and Tweedledum.

Albert Johannsen, 1932

Aa The native Hawaiian word for a lava flow that has a very rough and uneven surface, generally scoriaceous, with sharp, angular fragments or clinker. It is the international geological term for this type of lava flow and was first mentioned in print by Robert C. Haskell in 1859.

Andesite A volcanic rock with a chemical composition intermediate between that of basalt and rhyolite. The name was given by the German geologist Christian Leopold van Buch (1774–1853) to some volcanic rocks from the Andes Mountains in South America. The term was used by Charles Darwin in 1844.

Ash Scrope may have been the first to use this term in 1823 for fragmentary or pulverized volcanic materials, in connection with the eruption of Vesuvius in 1822. Bischof (1839) uses the term ash for pyroclastic deposit in the Brohl Valley in the Eifel district in Germany.

Basalt This term refers to volcanic rocks that are characterized by low silica content; it represents the most abundant volcanic rock type on the Earth, the Moon, and probably other planets of the solar system. On Earth the basalts make up most of the oceanic crust. The origin of the word is uncertain, but it may refer to the Greek word for touchstone, used in antiquity for testing coins. An Ethiopian origin has also been proposed, as a derivation from the word *bsalt*, to cook. Pliny the Elder refers to basalt in the first century A.D., and notes that statues are made from it, both in Rome and Egypt. Agricola introduced the word basalt into geology in 1546, in connection with his description of the volcanic rocks of the Schlossberg at Stolpen in Germany.

Bombs Fist-size or larger lumps of volcanic rock ejected during eruption. Bischof used the term in 1839 for pyroclastic deposit in the Brohl Valley in the Eifel district in Germany.

Breccia A rock consisting of large, angular blocks of volcanic fragments. The term was first used by Scrope in 1825.

Dike A vein of subvertical sheets of solidified magma, filling fractures in the Earth's crust. The term was first used by Hall in 1812. Scrope (1825) considers dikes formed both by injection from below and as a result of injection of lava from above, when a lava flow travels over a fractured terrain.

Eclogite A rock composed of the minerals pyroxene, omphacite, and pink garnet. It is a deep-seated rock, of high-pressure origin, that may be abundant in the Earth's mantle. The term was early on used by René Just Haüy (1743–1822).

Fallout The settling and deposition of particulate matter, such as tephra and volcanic aerosols, out of an eruption plume and onto the Earth's surface.

Flood basalts Laterally extensive deposits of basaltic lava flows, resulting from outpouring of vast volumes of magmas during fissure eruptions.

Fumarole Low-level and generally steady output of low-temperature gases such as sulfur, halogens, water vapor, and carbon dioxide from dormant volcanoes and geothermal fields in volcanic regions. Scrope (1825) and Bischof (1839) were the first to mention fumaroles as exhalations of aqueous vapor.

Geology The first to use the term geology or *geologia* for the study of the Earth was the Italian scholar Aldrovandus of Bologna (Ulisse Aldrovandi, 1522–1605) in 1603.

Hydrothermal Produced by or pertaining to the action of waters heated by contact with deep-seated hot rocks. Robert Bunsen first applied this term in 1847 to the heated solutions themselves as well as the deposits or rocks produced by the deposition of chemicals from the solutions.

Igneous rocks Rocks that have solidified from a molten condition (Latin: *ignis*, fire).

Lava The first use of the term lava in print may have been by Borelli (1670), when describing the studies of Francesco D'Arezzo of the Etna volcano. Hall may have been first to use it in English in 1812. Scrope (1825) gives a definition of *lava* that is essentially the same as the modern definition of *magma* and thus does not restrict it to surface flows, as is the case today. The population in the vicinity of Vesuvius in Italy has long referred to red-hot lava flows as "lave di fuoco," and mudflows as "lave d'aqua" or "di fango."

Lapilli Volcanic fragments in the millimeter-size range; first used in English by Scrope (lapillo; 1825). Bischof uses the term *rapilli* for pyroclastic deposit in the Brohl Valley in the Eifel district in Germany (1839).

Lithosphere The outer rigid layer of the Earth; first used by Chamberlain in 1909.

Obsidian The ancient term for volcanic glass formed by the rapid quenching of magma of silicic or rhyolitic composition. It was a term used by Pliny the Elder in the first century A.D. (*lapis obsidianus*), who referred to its occurrence in Ethiopia. Darwin used the term in print in 1844.

Oceanic ridges The elevated regions of the ocean floor where the plates are moving apart, creating new oceanic crust by submarine volcanism. The term was first used by Mallett in 1858.

Pahoe-hoe The Hawaiian term for a lava flow with a smooth surface, first introduced into print by Robert C. Haskell in 1859.

Phenocrysts Large crystals embedded in fine-grained volcanic rocks.

Plinian eruption An explosive volcanic eruption characterized by a high eruption column of ash, pumice, and volcanic gases that may extend tens of kilometers up into the Earth's atmosphere above the volcano. This style of eruption is termed *plinian* in honor of Pliny the Younger, the first to describe such an event during the Vesuvius eruption of A.D. 79.

Porphyritic The texture of volcanic rocks that have large phenocrysts set in a very fine grained or glassy groundmass. The word is derived from the greek *porphyra*, the gastropod that yields purple dye. This term was used to describe certain pink, granitic Egyptian ornamental rocks in antiquity that contain large crystals in a purplish groundmass. One of the first to use the term was Pliny the Elder in the first century A.D. to describe the Egyptian rocks. Charles Darwin used it in print when describing volcanic rocks in 1844.

Pozzolana The Italian term for a variety of partly consolidated volcanic ash deposits, commonly used in the manufacture of cement. Scrope (1825) refers to volcanic ash by this term.

Pumice A froth of volcanic glass that forms very vesicular and low-density bubble-rich material, generally light grey in color. The term was first used by Theophrastus in the third century B.C. and later by Pliny who refers to its occurrence in the volcanic islands of Melos and Nisyros in the Hellenic arc and on the Aeolian Islands. It has been proposed that the word is derived from the Greek *spuma*, foam. In later centuries Scrope (1825) defines *pomice*.

Rhyolite A volcanic rock with a high silica content. The term is derived from the Greek word for flow or torrent and was coined by Ferdinand von Richthofen in 1860. It is synonymous with the term *liparite* (after Lipari Island) used by Justus Roth (1887) and F. Zirkel.

Scoria Vesicular fragments of lava, a few centimeters in size; from the Greek word for dung, first used in print by Scrope (1825).

Solfatara A volcano that is in a state of dormancy, but is characterized by the emission of volcanic gases and precipitation of sulfurous deposits at the surface. The term was first used in print by Sir William Hamilton in 1771.

Subaerial and *subaqueous eruptions* Scrope was the first to make a distinction between eruptions that occur on the sea floor and on the surface, and to emphasize their different character, in 1825.

Sublimations Precipitates of chemicals from fumarolic gases; first used by Scrope in 1825.

Tephra The collective term for solid particles of silicate glass, resulting from quenching of magma, transported in eruption plumes in the atmosphere. These particles range in size from a few centimeters to microns (volcanic dust). The term was first proposed by the Icelandic geologist Sigurdur Thorarinsson in 1934.

Trachyte This term was early on given to a wide variety of volcanic rocks of acid or high-silica composition, many of which were later referred to as rhyolites. The first use of the term is by the French mineralogist René Just Haüy, in reference to volcanic rocks from the Drachenfels on the Rhine. The modern usage of the term is restricted to volcanic rocks of intermediate to high silica content, but with a high content of alkalis.

Trass A now obsolete term for fragmentary volcanic deposits. Scrope (1825) refers to *trass* in the Eifel as product of *igneo-aqueous* activity. Bischof (1839) also uses the term *trass* for tuff or pyroclastic deposit in the Brohl Valley in the Eifel district in Germany.

Tuff The Italian word "tuff" or "tufa" is synonymous with "ash" and commonly used for consolidated volcanic ash deposits.

Vesicles Air bubbles within a volcanic rock, produced by gases or volatiles that exsolve and expand out of the solidifying magma. The term was first used in print by Scrope (1825).

Volatiles Chemical elements or compounds soluble in magmas at high pressures and temperature that exsolve as gas from magmas during ascent and eruption at the Earth's surface. They include sulfur, water, carbon dioxide, chlorine, and fluorine.

Volcanic aerosol Very small (microns to a fraction of micron diameter) particles or droplets, composed mainly of sulfuric acid and water, produced by the conversion of sulfur dioxide gas to sulfuric acid particles in the volcanic eruption plume.

Xenolith A rock fragment, derived from deep-seated rocks or walls of the volcanic conduit, that is transported to the surface in magma and included in a lava flow. The term was first used by W. J. Sollas in 1894. *Xenos* is Greek for a guest or stranger.

Index